T0247847

# ADRIFT

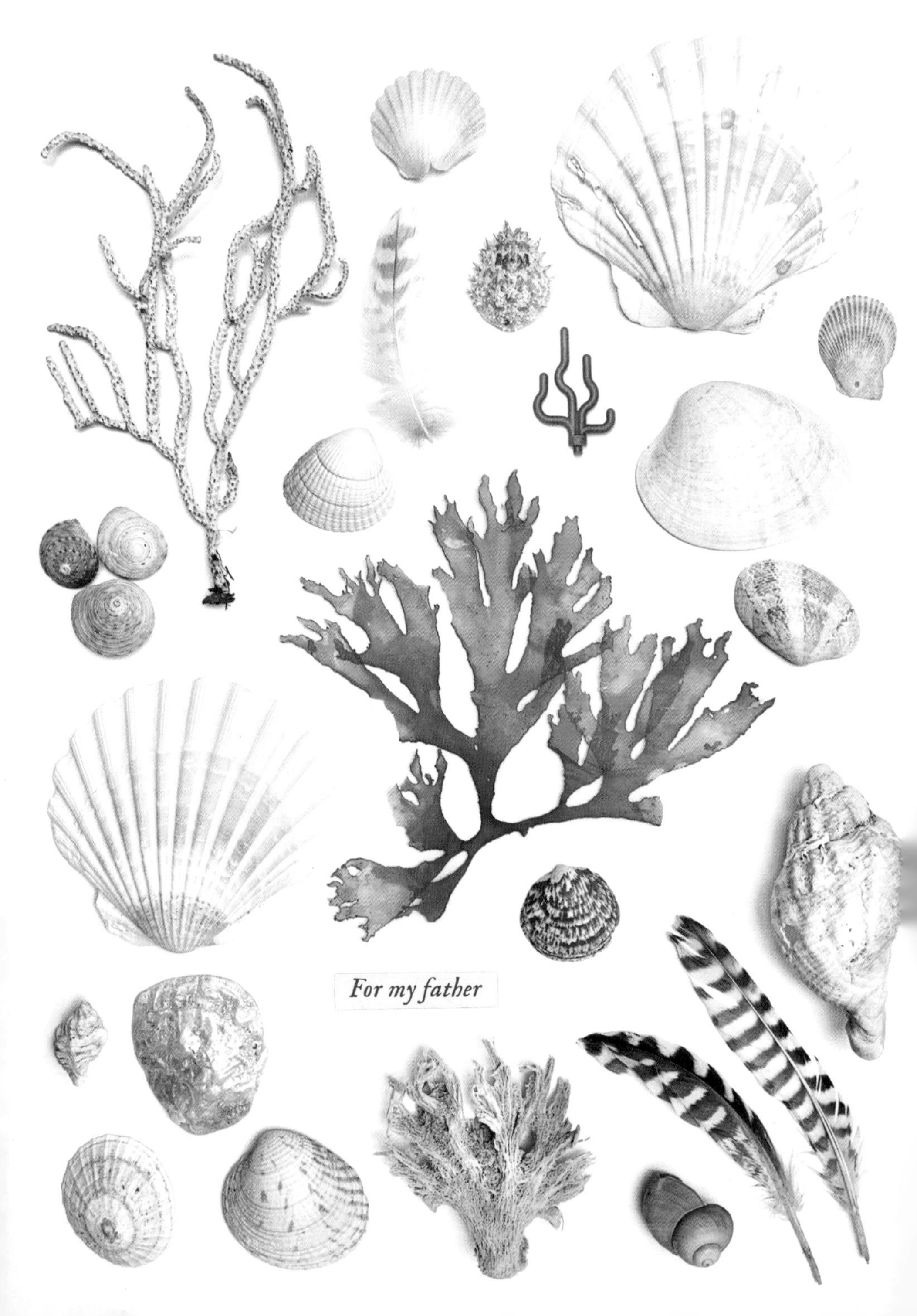

*For my father*

# ADRIFT

## THE CURIOUS TALE OF THE LEGO LOST AT SEA

*A voyage into the changing nature of beachcombing, dragon hunting, plastic in the sea and sand, and the lasting legacy of a cargo spill.*

TRACEY WILLIAMS

*with Dr Curtis Ebbesmeyer and Mario Cacciottolo*

UNICORN

*Not just Lego – some of the thousands of pieces of plastic picked up from beaches after winter storms*

# CONTENTS

## *Author's Note*

I love Lego. I played with it as a child. My own children played with it. We still have boxes full of Lego in the attic, ready to be handed down to future generations. Searching for lost Lego from a cargo spill began as a bit of fun, a treasure hunt with my children. Ultimately, it opened my eyes to all the rest of the plastic in the ocean.

If you search carefully along the strandline after a wild winter storm, you might still find them. Tiny yellow life jackets and grey scuba tanks. Bright green plastic sea grass and little spear guns in red and yellow. Blue, black and red divers' flippers and miniature cutlasses. Perhaps a dragon or an octopus, just 3 inches long. Maybe even a small yellow life raft. They're from an armada of Lego that fell off a ship in 1997. And they're still turning up today.

When my children were young, searching for Lego on the shores beneath our family home on the south coast of Devon became the highlight of any trip to the beach. Every weekend would involve a treasure hunt. A cork, a feather, two Lego flippers, a heart-shaped stone.

It was a tradition that stemmed back to my own childhood in the 1960s, when my parents would pen lists of things to find on family holidays in Cornwall and we'd fossick for shells and sea glass among the sand and shingle.

I still remember the day a neighbour found a green Lego dragon. Even today, some twenty-five years later, she signs her Christmas card to me 'Mary, keeper of the green dragon'.

But the holy grail of each beachcombing trip was a black Lego octopus. We did eventually find one. It was in a remote cove in South Devon, its arms tangled in seaweed.

## Sea-Fever

JOHN MASEFIELD

I must down to the seas again,
to the lonely sea and the sky,
And all I ask is a tall ship and a star
to steer her by,
And the wheel's kick and the wind's
song and the white sail's shaking,
And a grey mist on the sea's face,
and a grey dawn breaking.

I must down to the seas again,
for the call of the running tide
Is a wild call and a clear call that may
not be denied;
And all I ask is a windy day with the
white clouds flying,
And the flung spray and the blown
spume, and the sea-gulls crying.

I must down to the seas again,
to the vagrant gypsy life,
To the gull's way and the whale's way,
where the wind's like a whetted knife;
And all I ask is a merry yarn from a
laughing fellow-rover,
And quiet sleep and a sweet dream
when the long trick's over.

Things to Find

cork
feather
cockle shell
Blue sea glass
Two Lego flippers
mermaid's purse
sea potato
Lego pirate's cutlass
heart-shaped stone
black Lego dragon

# 1 Lego Galore

## *How millions of Lego pieces ended up in the sea*

The curious tale of the Lego lost at sea began on Thursday 13 February 1997, when a cargo ship laden with goods was hit by a storm.

The *Tokio Express* had set sail from Rotterdam in the Netherlands when it became engulfed in mountainous seas 20 miles off Land's End, Cornwall.

In what the ship's captain later described as a 'once in a 100-year phenomenon', a rogue wave tilted the vessel 60 degrees one way, then 40 degrees back, sending sixty-two shipping containers toppling into the ocean.

One held nearly 5 million pieces of Lego, on its way from the toy company's factory in Billund, Denmark to North America, where it was to be made up into sets.

By a strange quirk of fate, much of the Lego was sea-themed; 4,756,940 pieces of plastic, bound for seafaring adventures.

No one is sure what happened next – whether the container of Lego burst open on impact, scattering its contents into the ocean, or whether it carried on floating for a while, slowly releasing its cargo as it drifted to the seabed.

But in the days that followed, helicopter pilots flying over the area reported seeing 'a slick of Lego' floating in the sea. And beachgoers started finding Lego washed up on Cornwall's wild and windswept shores.

# Beachcomber

GEORGE MACKAY BROWN

Monday I found a boot –
Rust and salt leather.
I gave it back to the sea, to dance in.

Tuesday a spar of timber worth thirty bob.
Next winter
It will be a chair, a coffin, a bed.

Wednesday a half can of Swedish spirits.
I tilted my head.
The shore was cold with mermaids and angels.

Thursday I got nothing, seaweed,
A whale bone,
Wet feet and a loud cough.

Friday I held a seaman's skull,
Sand spilling from it
The way time is told on kirkyard stones.

Saturday a barrel of sodden oranges.
A Spanish ship
Was wrecked last month at The Kame.

Sunday, for fear of the elders,
I sit on my bum.
What's heaven?
A sea chest with a thousand gold coins.

# LEGO BONANZA FOR BEACH-GOERS

**BEACHCOMBERS** are searching the coastline for Lego after a shipping crate filled with millions of colourful pieces fell from a ship during a storm 20 miles off Land's End, the most westerly point of mainland England. The bus sized container was one of sixty-two lost overboard from the cargo vessel *Tokio Express* on 13 February 1997 when it was hit by a freak wave during severe gales. Inside the container were nearly 4.8 million pieces of Lego, sorted into red, green, blue and yellow boxes. Many were sea-themed.

The Lego was on its way from the toy company's headquarters in Denmark to North America, where it was to be made up into kits.

Wheelbarrow wheels, beer, hose parts, garden tools, furniture, French perfume, car parts and cigarette lighters were reportedly in some of the other containers lost overboard.

The stretch of water where the shipping containers toppled like dominoes from the *Tokio Express* is treacherous. Over the centuries, many vessels have foundered here. From ancient wooden sailing ships to a German U-boat destroyed in 1945 with the loss of her crew, the shifting sands conceal dozens of wrecks. Beneath the waves, too, lie the remains of cargo ships torpedoed in the First World War, their holds once laden with salt and copper ore.

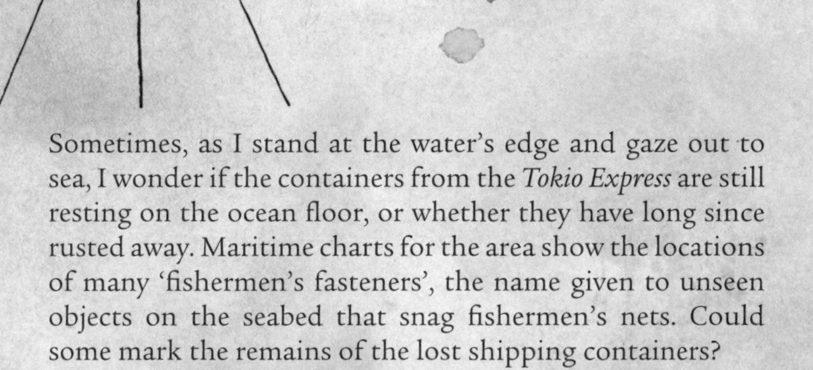


Sometimes, as I stand at the water's edge and gaze out to sea, I wonder if the containers from the *Tokio Express* are still resting on the ocean floor, or whether they have long since rusted away. Maritime charts for the area show the locations of many 'fishermen's fasteners', the name given to unseen objects on the seabed that snag fishermen's nets. Could some mark the remains of the lost shipping containers?

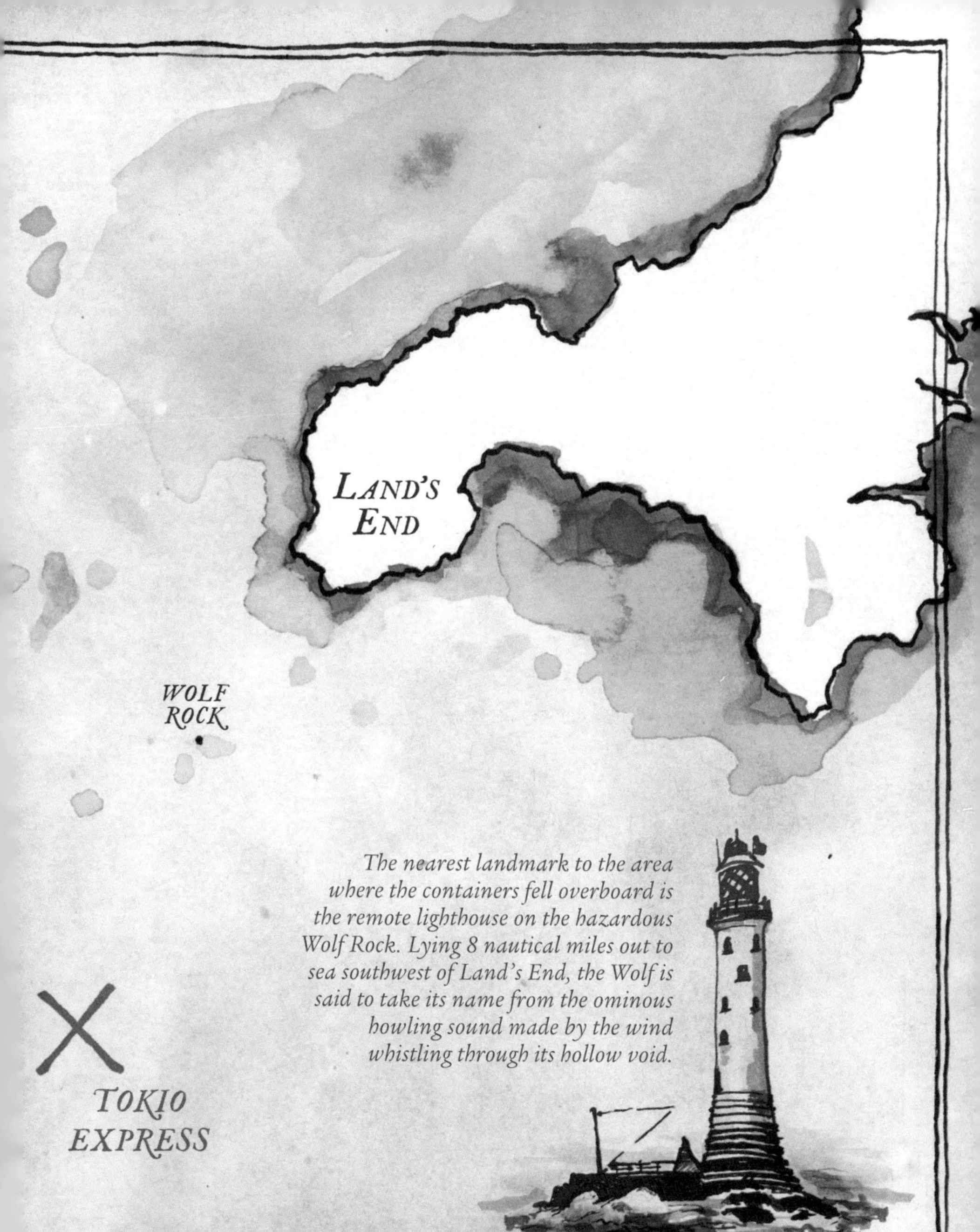


*The nearest landmark to the area where the containers fell overboard is the remote lighthouse on the hazardous Wolf Rock. Lying 8 nautical miles out to sea southwest of Land's End, the Wolf is said to take its name from the ominous howling sound made by the wind whistling through its hollow void.*

The story of the Lego lost at sea began for me not in Cornwall but in the neighbouring county of Devon.

In the early 1980s, my parents had moved to an old house on the south coast known as 'The Eyrie' or 'Eagle's Nest', perched high on the cliffs overlooking the sea. From there we'd watch wild storms blowing in from the Atlantic, sending huge rollers crashing on to the rocks beneath the house.

'The sea will provide,' my father would tell us, as the skies turned black and the waves thundered in. And it was there, on the shores below our home on the cliff, that we first started to find the Lego.

We noticed the smaller pieces first. Tiny life jackets strewn along the strandline, spear guns and scuba tanks scattered across the sand, flippers and flowers floating in rock pools. We gathered the pieces up in their hundreds, storing them at first in old ice cream tubs on the kitchen windowsill.

Then, as our hoard grew, we transferred them to big plastic tubs that my father had once used for brewing beer in our old wooden shed, which was lashed to the clifftop with pegs and guy ropes to stop it blowing away in a gale.

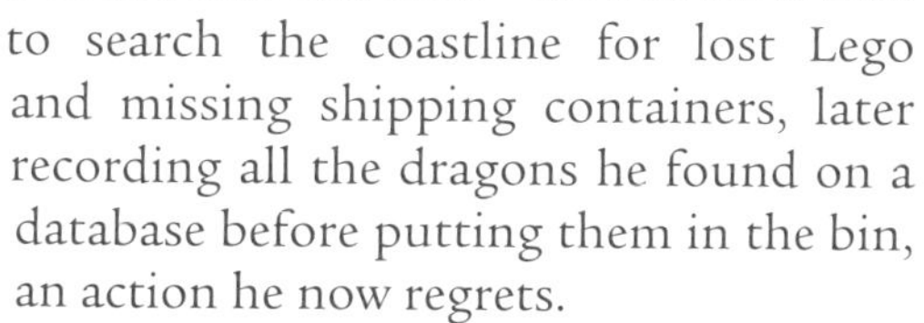

It was the dragons that captured everyone's imagination. There were 33,941 inside the container that fell off the *Tokio Express* – 33,427 black dragons and 514 green. While the black dragons washed ashore in their thousands, the green ones proved to be far more elusive, with very few ever reported.

Only the dragon bodies made it to land, though. All were missing their arms, tails, upper jaws and bright orange wings, which led to some confusion as to what they were. One beachcomber later recalled finding Lego dinosaurs washed up, while another spoke of Lego seahorses.

Tales emerged of children filling buckets with dragons and selling them at car boot sales for 10p each. One beachcomber described how her mother had made her rummage through mounds of rotting seaweed for weeks, desperate to find one. Council vehicles were said to have mechanically raked them up from the strandline.

A former member of the coastguard told how – as a new recruit on one of his first call outs – he had been sent to search the coastline for lost Lego and missing shipping containers, later recording all the dragons he found on a database before putting them in the bin, an action he now regrets.

Margaret Jones took up dragon hunting in her mid-eighties, eventually finding twelve. She discovered all her Lego dragons on the south coast of Cornwall in the late 1990s, often taking her grandchildren along to search. 'Mum loved the beach and beachcombing,' says

*Dragon hunter Margaret on Polzeath Beach in 1999 when she was 87. She lived to a happy 98.*

The Polkerris haul, which included nine dragons

A FAMILY DIARY ENTRY FOR 27.03.97 READS

*'Polkerris – Menabilly. Found lots of Lego on the beach.'*

her daughter Gwynneth. 'She shared her dragons around the family. I still have a green dragon and a black one living on a bookshelf in Norfolk.'

Another daughter, Ruth, also has vivid memories of dragon hunting. 'I remember my mother finding numerous dragons on Portwrinkle Beach in Cornwall in 1997,' she recalls. 'My husband recorded the Lego in his diary. An entry for 27.03.97 reads: "Polkerris – Menabilly. Found lots of Lego on the beach", while another for 11 August 1997 reads "Polkerris – found lots of Lego including flippers and helmets."'

After a hunt through their garage, the 'Polkerris haul' was found, including eight black dragons and one green one.

*Dave Smethurst*

Beachcomber Dave Smethurst found two Lego dragons during his regular, extensive beach cleans. In a six-month period, he single-handedly removed 48,947 bits of plastic from beaches around Cornwall, carefully counting and recording it all.

Sadly, Dave died in 2016, but every year on the anniversary of his death, his children and grandchildren do a beach clean in his memory.

'I still remember the storm,' recalls Mary, who was beachcombing with her two-year-old grandson when they discovered an elusive green Lego dragon washed up at Bigbury Bay on the south coast of Devon. 'We started finding the Lego soon afterwards. There were flippers, scuba tanks, spear guns and sea grass everywhere.

'The dragons were the real find. We only found a few of them. My grandchildren loved to hear the story. We told them how the dragons had been floating in the sea and how happy they were to have made it on to dry land.'

*Dragons found by Suki Honey*

*Did you know that a collective name for dragons is a flight?*

You can also have...

- a *thunder* of dragons
- a *blaze* of dragons
- a *horde* of dragons
- a *drive* of dragons
- a *weyr* of dragons
- a *wing* of dragons
- a *doom* of dragons

Of all the Lego that washed ashore, the black Lego octopuses were the most coveted, often being described as the holy grail of finds from the spill.

Although there were tens of thousands of dragons on the *Tokio Express*, there were only 4,200 octopuses, making them highly prized among beachcombers.

It is said that octopuses are masters of disguise, blending into the background to escape detection. Lego octopuses are no different. When tangled in seaweed, they can be almost impossible to spot. I found my first Lego octopus back in 1997 but didn't discover another for eighteen years.

*Lego octopus tangled in seaweed – would you have spotted it? This one was found by Suki Honey in southeast Cornwall.*

# LEGO TO LOOK OUT FOR...

Some bits of Lego from the spill turn up more regularly than others. These are the pieces beachcombers find most often. *The numbers show the quantities lost overboard rather than those found by beachcombers.*

## 88,316
### SETS OF FLOWERS

These originally came in sets of four, with each flower attached to a central sprue, or doodah, as beachcombers call them. In the container were 32,000 white sets, 40,316 red and 16,000 yellow. If they had separated at sea there would have been 441,580 flowers and sprues adrift, although many sets washed up intact.

## 54,000
### PIECES OF SEA GRASS

These were destined for Lego sets such as Sting Ray Explorer, Shark Cage Cove and Discovery Station. Easy to recognise but harder to find these days.

## 97,500
### SCUBA TANKS

'Old grey' scuba tanks featured in many ocean-themed Lego sets from 1997, including Diving Expedition Explorer, Deep Reef Refuge and Discovery Station. Still a regular strandline find.

## 79,680
### SPEAR GUNS

Lost overboard were 53,120 yellow spear guns and 26,560 red. While yellow spear guns featured in 1997 Lego sets, such as Deep Reef Refuge and Shark Attack, red were included in Deep Sea Bounty and Shark Cage Cove.

## 8,100
### GREY LINKS

Not the most exciting of pieces as far as beachcombers are concerned, these 'old grey' links featured in Discovery Station, a Lego Divers set released in 1997.

## 26,600

### LIFE JACKETS

Bright yellow life jackets came in quite a few Lego sets from 1997, including Scuba Squad and Deep Sea Bounty.

## 50,000

### BROOMS

These were included in sets such as the Witch's Magic Manor, Night Lord's Castle and Witch's Windship. When they're mingled in with seaweed and driftwood on the strandline, these little brown broomsticks are difficult to spot.

## 26,400

### BITS OF SHIP RIGGING

Measuring 5 inches long, these came in several Lego sets from 1997, including Pirates Perilous Pitfall where they were used to create the shipwreck and the Witch's Windship, where they formed part of the gondola pulled by the green dragon.

## 92,400

### CUTLASSES OR SWORDS

A favourite find, these featured in various Lego sets from 1997, including Pirates Perilous Pitfall and the pirate ship Cross Bone Clipper. Once we found a Lego cutlass in a tadpole-filled pond in the dunes, carried there on the crest of a wave during a storm.

## 352,000

### PAIRS OF FLIPPERS

These divers' fins or flippers were shipped in pairs: 209,000 black, 121,000 blue and 22,000 red. If all the pairs had separated at sea, there would have been 1,056,000 pieces of plastic adrift: 704,000 flippers or fins and 352,000 sprues or doodahs. Not all broke apart, however.

# 2

# Carried in Waves

## *The lost Lego looks for land*

Adrift! A little boat adrift!
And night is coming down!
Will no one guide a little boat
Unto the nearest town?

So Sailors say – on yesterday –
Just as the dusk was brown
One little boat gave up its strife
And gurgled down and down.

extract from **Adrift**
EMILY ELIZABETH DICKINSON

*The arrival of beautiful 'by-the-wind sailors' or* Velella velella *on beaches is believed by many to be a sign that flotsam is on its way. Not a single creature but a colony of tiny individual animals, by-the-wind sailors drift on the surface of the ocean at the mercy of wind and currents. With their vivid blue colour and tiny sails running diagonally across their body, they're easy to recognise. It's often said that most by-the-wind sailors are left-handed, their sails running from upper left to lower right. But, once, we found thousands of right-handers washed up.*

News of the Lego spill soon reached Dr Curtis Ebbesmeyer, the pioneering oceanographer famed for unravelling the mystery of ocean currents.

He first captured the world's imagination when he called on a network of beachcombers to track the passage of nearly 80,000 Nike shoes lost overboard from the cargo vessel *Hansa Carrier* in May 1990, while on its way from Korea to the United States. As each of the shoes carried a unique serial number, he was able to work out which had come from the spill, gaining a valuable insight into ocean currents.

Two years later, Dr Ebbesmeyer called on those same beachcombers to help him track the journey of 28,800 plastic bath toys from a cargo spill in the Pacific Ocean, as they washed up around the world. The 'Friendly Floatees' as they were known – yellow ducks, blue turtles, red beavers and green frogs – had fallen overboard from a container ship in 1992, while en route from Hong Kong to Tacoma, Washington. They were subsequently discovered on beaches from Alaska to Scotland and Texas.

Dr Ebbesmeyer thought the toys could in time become frozen in Arctic ice, eventually being carried by ocean currents to Europe and Florida as the ice melted and thawed into the Atlantic Ocean.

Although quite a few people claim to have found plastic ducks from the spill, there are many impostors. The true Friendly Floatees ducks have 'The First Years' embossed on their tiny chests.

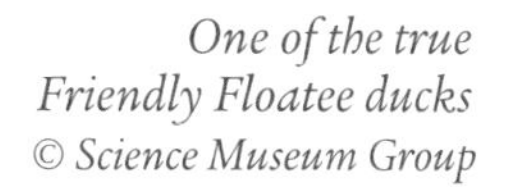

*One of the true Friendly Floatee ducks © Science Museum Group*

When he heard about the Lego washing up on beaches in England, Dr Ebbesmeyer contacted the toy company's headquarters in Denmark to ask what had been in the container. In response, Lego's shipping department sent him a box of samples, along with an inventory. It not only revealed which Lego bricks had been on board the *Tokio Express* but also how many of each piece had plummeted into the deep. It was, he said, the most detailed list of a lost shipping container's contents he had ever obtained.

*'Imposter' ducks found washed up on Cornish beaches*

# WAVE WATCHING

*Oceanographer Dr Curtis Ebbesmeyer, who began following the Lego spill back in 1997, explains how he tracked a lost cargo of plastic ducks*

'To paraphrase the American poet Edwin Arlington Robinson, the ocean is forever asking questions and writing them on the shore.

'So when you see something on the beach, it's asking you something. Your job is to find out the answer.

'I've been fascinated with the sea all my life, even as a child. When I grew up, I studied oceanography at the University of Washington, got my PhD in 1973 and have been charting flotsam (things that fell overboard) and jetsam (things thrown overboard on purpose) ever since.

'Most people know the Earth's land mass is made up of tectonic plates, but that's the solid part of the planet. It's also covered by fluid plates, huge masses of water called gyres, which circulate around our seas. That's why we have what I call a 'clockwork ocean' made up of many moving parts.

'Think of the gyres as the face of a clock. The water moves around the face, and the numbers are the points where items floating in it might end up. Different gyres take different lengths of time to complete their pattern of moving around. It can take anything from nine months to six years, depending on which gyre you're talking about. I discuss these gyres in more detail in my book *Flotsametrics and the Floating World*.

'The most profound lesson I've learned from the Lego story is that things that go to the bottom of the sea don't always stay there, even when they're in a steel container. They can still be carried around the world, subject to the planet's currents and tides.

'Tracking currents is like tracking ghosts. You can't see them; all you have to work with is where flotsam starts and where it ends up.

'In 1992 a ship called the *Ever Laurel* lost a container overboard, which resulted in 7,200 red beavers, 7,200 green frogs, 7,200 blue turtles and 7,200 yellow ducks – 28,800 bathtub toys in total – being released into the Pacific Ocean.

'I conducted a great deal of research on this, carrying out computer modelling with my colleague Jim Ingraham, and we managed to track the floating armada as it bobbed its way around the high seas.

'From the site of the spill in the Pacific, they floated across the International Date Line, up past the North Pole, down past Greenland and the UK, then down to Spain and across the Atlantic, where some washed up in the Gulf of Mexico, in Texas.

*'The most profound lesson I've learned from the Lego story is that things that go to the bottom of the ocean don't always stay there.'*

Dr Curtis Ebbesmeyer

'The ducks got a lot of attention – I mean, everyone loves a plastic yellow duck – but I like to tell people how one green frog washed up in the Gulf of Mexico, having been in the sea for twenty-six years.

'That part of the United States – the Great Bend of Texas – gets a lot of items washed up on it. The ocean is very tidy. She likes to put the things she carries in their place.

'I'm often asked how long plastic can survive in the sea. Well, in 1944 a military aircraft crashed into the ocean near the Philippines, and sixty years later a piece of plastic the size of a thumbnail from the plane was discovered in the stomach of a dead albatross chick, which is a sad reality of what plastic pollution can do to wildlife.

'It looks like James Bond had it wrong. It's not diamonds that are forever. It's actually plastic.'

## Jetsam

Goods cast overboard to lighten a vessel in danger of sinking. The vessel may still perish.

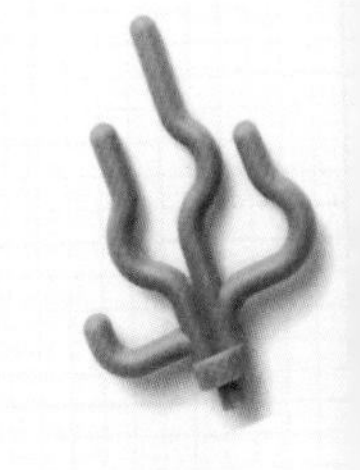

## Flotsam

Goods lost from a ship which has sunk or otherwise perished. Goods are recoverable because they remain afloat (for example, Lego from the *Tokio Express* that floated).

## Lagan

Goods cast overboard from a ship which afterwards perishes. The goods are buoyed so they can be recovered.

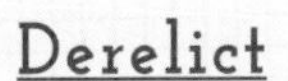

## Derelict

Property, whether vessel or cargo, which has been abandoned and deserted at sea by those who were in charge of it without any hope of recovering it (for example, Lego from the *Tokio Express* that sank).

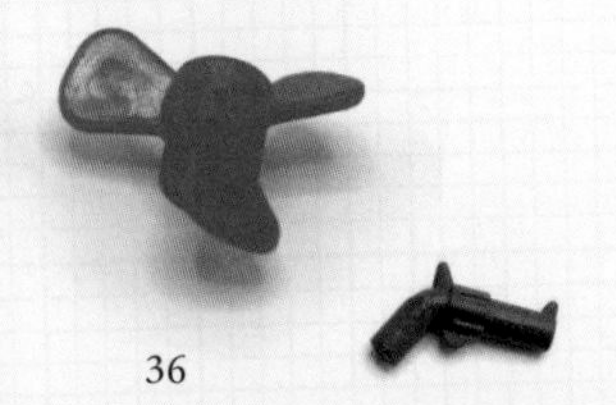

Inside the box of samples sent to Dr Ebbesmeyer were one hundred different pieces of Lego, many of them sea-themed. Stuck to each was a tiny paper label with a corresponding part number, handwritten in ink.

Curious to find out which of these might have been cast adrift and which may have sunk to the bottom of the ocean, Dr Ebbesmeyer put them all into water, later separating them into those that floated and those that did not.

'I used fresh water, reasoning if they floated in that they would be quite buoyant in seawater,' he explained, adding he did the experiment quickly because he didn't want the little paper labels to come off.

Writing in the summer 1997 edition of his newsletter *Beachcombers Alert!*, Dr Ebbesmeyer predicted beachcombers would find pieces from at least ten Lego themes, so long as they were adrift: Divers, Aquazone, Aquanauts, Police, Fright Knights, Wild West, Roboforce, Time Cruisers, Outback and Pirates.

Among the pieces to look out for, he said, were yellow life rafts, dragons, octopuses, sea grass, life jackets, spear guns, scuba tanks, frying pans, black bats, witches' hats, cowboy hats, divers' flippers, flowers and monster feet.

In total, he calculated over 3 million bits of Lego might be afloat, assuming the sets of Lego flowers and pairs of flippers separated at sea, although not all of them did.

Some, he predicted, would reach the USA by the summer of 1998 and would probably wash up along the coast of Florida, Georgia and the Carolinas.

'Remember flotsam, like birds of a feather, flocks together,' he wrote. 'If you find a yellow raft, look around

for the smallest elements such as daisies, swords, pistols, hats, flippers and spear guns.'

Dr Ebbesmeyer thought the size of some of the pieces would challenge beachcombers, though. 'The yellow rafts, because of their brightness and number (28,700 lost) may likely be the most oft reported,' he said.

*'Remember that flotsam, like birds of a feather, flocks together.'*

Dr Curtis Ebbesbeyer

But strangely, sightings of the Lego life rafts proved to be few and far between.

'Any week now, a bright yellow life raft should wash ashore somewhere between Miami Beach and Cape Canaveral, ending an epic voyage spanning more than a year and thousands of nautical miles,' wrote journalist Curtis Morgan in an article about the Lego spill in the US newspaper, the *Miami Herald*, the following year.

Whether any of the life rafts made the epic voyage from the UK to the US, no one really knows. Although some oceanographers believe the boats could have circled the Atlantic by now, very few have been reported.

Every year, the organisers of the International Sea-Bean Symposium and Beachcombers' Festival in the US would offer a prize for the first beachcomber to find a piece of Lego from the spill.

However, year after year, the prize went unclaimed. 'I go to beachcombing events in Florida and they show me Lego – but it's the wrong kind,' Dr Ebbesmeyer said later in an interview with the BBC. 'It's all local stuff kids have left behind.'

Back in the UK, beachcombers were still finding the same pieces of Lego washing up. Life jackets, scuba tanks, flippers. Flowers, witches' brooms, sea grass. Ship rigging, spear guns, cutlasses. The much sought-after dragons, the occasional octopus. And the less interesting grey link or axle, that no one realised was Lego until one sharp-eyed beachcomber eventually spotted the logo on it.

Where were all the witches' hats and black bats? Where were the pistols and frying pans? Where were all the monster feet?

Over the years, I largely forgot about the Lego. My children were older, we were living inland, hundreds of miles from the sea, and our trips to the beach became less frequent. But in 2010, I moved to the north coast of Cornwall, to be closer to family. And on my first visit to the beach, I found a bright yellow Lego life jacket from the spill. Thirteen years on and it was still turning up. I was amazed.

There wasn't just Lego on the strandline, though. There was so much plastic. Shoes, buoys, rope, bottles. Some from shipping, some from fishing, some left behind by beachgoers. But where did all the rest come from? The toothbrushes, the bookies' pens, the paintbrushes and monopoly houses?

Horrified at the amount of plastic now littering the beach, I joined the barefoot army of dogwalkers, surfers and beachgoers picking up debris from Cornwall's shores. I began working round the tides, often leaving the house before dawn to roam the coastline with my rescue dog, Jess.

While mudlarks like to visit the river Thames in London at low water when more of the foreshore is exposed, I would search along the strandline at high water to see what the ebbing tide had left behind. Whereas once I'd sought out shells and sea glass on family holidays in Cornwall as a child, now I hunted for plastic, filling bags with debris cast up by the sea.

Meanwhile, in the US beachcombers were still on the lookout for Lego. 'Scientists anticipate thousands of colourful little Lego toy sea creatures to wash up on our shores,' ran a poster for the Seventeenth Annual Sea-Bean Symposium and Beachcombers' Festival in 2012. 'Florida beachcombers are eager to be the first in America to sight them. Oceanographers around the world, as well as publications such as *National Geographic*, are waiting to hear about the Lego armada. Will they finally be here in 2012–2013?'

Then, in July 2013, a winner was announced. A beachcomber in the USA had discovered a black Lego octopus tangled in thick, dried sea grass on Galveston Island, Texas. Could it have been from the spill? It certainly matched the description of the 4,200 octopuses that fell into the ocean in 1997.

*'Scientists anticipate thousands of colourful little Lego toy sea creatures to wash up on our shores. Florida beachcombers are eager to be the first in America to sight them.'*

SEVENTEENTH ANNUAL SEA BEAN SYMPOSIUM AND BEACHCOMBERS' FESTIVAL, 2012

Back in the UK, the same bits of Lego were still turning up. 'I am beginning to think that perhaps a cargo container rests on the sea floor, slowing discharging Lego as it disintegrates,' wrote Dr Ebbesmeyer in September 2013.

'Perhaps the door is slightly ajar. The metal in a container's walls is ⅜-inch thick, so it may take some years for the container to disintegrate sufficiently to let all of its Lego go.'

Curious to discover who else was finding the Lego, what pieces they had picked up and how far it had drifted in the intervening years, I set up a Facebook page in October 2013 called 'Lego Lost At Sea'. I thought the information it generated might be useful to oceanographers studying movements of the ocean, as well as to scientists investigating plastic in the sea.

Then in June 2014, a message arrived from a journalist at the BBC. He was interested in doing a story about the lost Lego, planned to visit Cornwall and wanted to meet.

# LANDING THE LEGO STORY

*Mario Cacciottolo, the journalist whose BBC article about the Lego found on beaches went viral, explains how he learned about the cargo spill and which piece he found in the sands himself.*

'I was half-way through a year-long trip around the world between 2013 and 2014 when I spotted a single sentence in a Hong Kong newspaper story that I will never, ever forget. It was a quote from the oceanographer Dr Curtis Ebbesmeyer, who said pieces of Lego were still washing up daily on a beach in Cornwall, because of a container spill many years before.

'I continued to drag my over-stuffed backpack across far-flung lands and seas for many more months during my sabbatical year off from the BBC News website, but while I threw away some things to lighten my load, I never let go of that piece of tantalising trivia. I began to research to see if this story was true. To my delight I found it was.

'So, when I eventually returned to work at the BBC's newsroom in central London early in the summer of 2014, literally one of the very first things I did was write an email pitch to my editor, asking him to commission the story. I remember punching the air in delight at my desk when he replied with one word: "Yes."

'I started building up the story but I needed more, someone to click these pieces of plastic together and build up the full picture. Luckily my digging around led me to the "Lego Lost at Sea" page on Facebook and its creator Tracey, who shared many of the things she'd learned about the *Tokio Express* and its sunken secrets.

'I flew to Cornwall to meet her, and we walked along Perranporth Beach while the cloudless sky bashed us with hot sunshine. We did a photoshoot with the dragons and octopuses she'd previously plucked from the sand's clammy embrace (made mildly complicated by the elderly nudists who kept wandering into the background).

'Then, as we dug around the seaweed and the rocks, I found a tiny white plastic flower, a mere hint of the cargo spill all those years ago. I popped it into my wallet and flew back to London, convinced this story would prove popular. It turned out to be an absolute whopping smash, breaking the dials and knobs which record the BBC's news website traffic, and becoming a story read by millions of people around the world.

'I was fully aware, and included in my article, that this tale is one of how our seas are swirling and swishing with things they should never have swallowed. But despite this sobering fact, I know that of all the many topics I wrote about as a journalist, the story of the lost Lego pieces is the one which brought me the most joy.'

After the BBC story, beachcombers came forward in their hundreds to share stories and memories of their Lego finds, many posting images on social media. It seemed the Lego had drifted far and wide. One beachcomber recalled finding the battered and broken lid of a Lego Group box among piles of seaweed washed ashore on the south coast of Cornwall, later taking it home by dinghy to live in her garden.

Pictures emerged of dragons and octopuses perched on bookshelves, lined up for photos, reunited for family gatherings and woven into tapestries. There were even reports from as far afield as Portugal and Australia, where beachcombers said they had discovered cutlasses, flowers and flippers.

The Facebook page became a joyous affair, with people sharing videos of the 'happy dances' they did when they found a bit of Lego. At Whitsand Bay in Cornwall, beach cleaner Michelle was so overwhelmed when her daughter found a Lego octopus, she feared she might faint, falling to the ground with her head between her knees.

As the BBC article, and all the other reports that came after it, continued to get huge attention around the world, people travelled vast distances to hunt, flying in from the USA, Italy, Switzerland and Belgium.

When Martin and his son travelled from the Netherlands to Cornwall in their quest to find a dragon, they brought their own Lego dragons with them to help with the search. Although a dragon from the *Tokio Express* proved elusive, they did discover a piece of rigging from the Witch's Windship, later taking it back home with them.

In December 2014, we took a film crew on a Lego hunt for the documentary *The Secret World of Lego*, which went behind the doors at Lego's HQ. The story was even turned

into a stage production by theatre-makers. In 2017, *Lego Beach* made its debut at the Bristol Old Vic, with sea shanties played on the accordion 'to create a stormy, tidal atmosphere', said the producers.

Tour companies were keen to lay on Lego hunting trips, an opportunity beachcombers declined, fearing a backlash if the much-anticipated dragon failed to appear. Later in the year, the Lego spill featured in an exhibition named *Container* at the Australian National Maritime Museum about the humble shipping container, the box that changed the world.

There were tales, too, of the Lego that got away.

The fisherman who moved to a new home in the ancient Cornish fishing port of Newlyn, forgetting he had Blu-Tacked a Lego octopus to his wall in Penzance.

Environmental campaigner and marine debris artist Rob Arnold, who found a Lego octopus while being filmed by a Finnish TV crew for a documentary about beachcombing, then promptly lost it again, a moment also captured on film.

And author and photographer Lisa Woollett, who threw out the old sea-worn dragon she found, not realising its 'treasure-significance'. As luck would have it, she did find another many years later. You can see it on the front cover of the paperback edition of her book, *Rag & Bone: A History of What We've Thrown Away*.

*Lisa's dragon*

# A LEGO ODYSSEY

To find out how far the Lego had drifted, a friend and I began recording where and when it washed up.

It seemed much had been swept by ocean currents up the north coast of Cornwall, where it was discovered on almost every beach from Land's End to Bude, with Perranporth proving to be a hotspot for Lego brooms.

*WALES*

From there the Lego flotilla headed up the North Devon coast, before reaching Wales, where a black Lego dragon washed ashore at Porthcawl. More sightings were made on the Gower Peninsula, then in Pembrokeshire and North Wales.

The reports led to further coverage by Mario at the BBC News website as we worked together to chart the discoveries. His second story on the Lego spill, this time with maps of the find locations, was published in January 2015.

And the Lego finds just kept on coming. In 2018, a piece of Lego sea grass was found in Cumbria at St Bees and given a local BBC news story all of its own.

There were sightings on the Isles of Scilly where one beachgoer recalled finding so many dragons he buried most in the dunes, hiding more in 'little dragon caves' for others to discover.

*ISLES OF SCILLY*

In Ireland, a Lego octopus was found at the foot of a cliff at Waterville, a seafront village in County Kerry. Then, in early 2021, a dragon was spotted at Spanish Point in County Clare.

*WALES*

Meanwhile, thousands of Lego pieces had been scattered along Cornwall's south coast, with numerous sightings of Lego octopuses on the Lizard Peninsula. More octopuses were seen at Whitsand Bay, a known dragon hot spot, where flippers galore turned up too.

In South Devon, dragons and octopuses made landfall at Bigbury Bay, while in Dorset more castaway Lego drifted ashore, including a dragon at Chesil Beach and an octopus at Worbarrow Bay.

From there, the Lego fleet made its way along the coast of Hampshire before reaching Sussex, where flippers

*IRELAND*

IRELAND
GUERNSEY

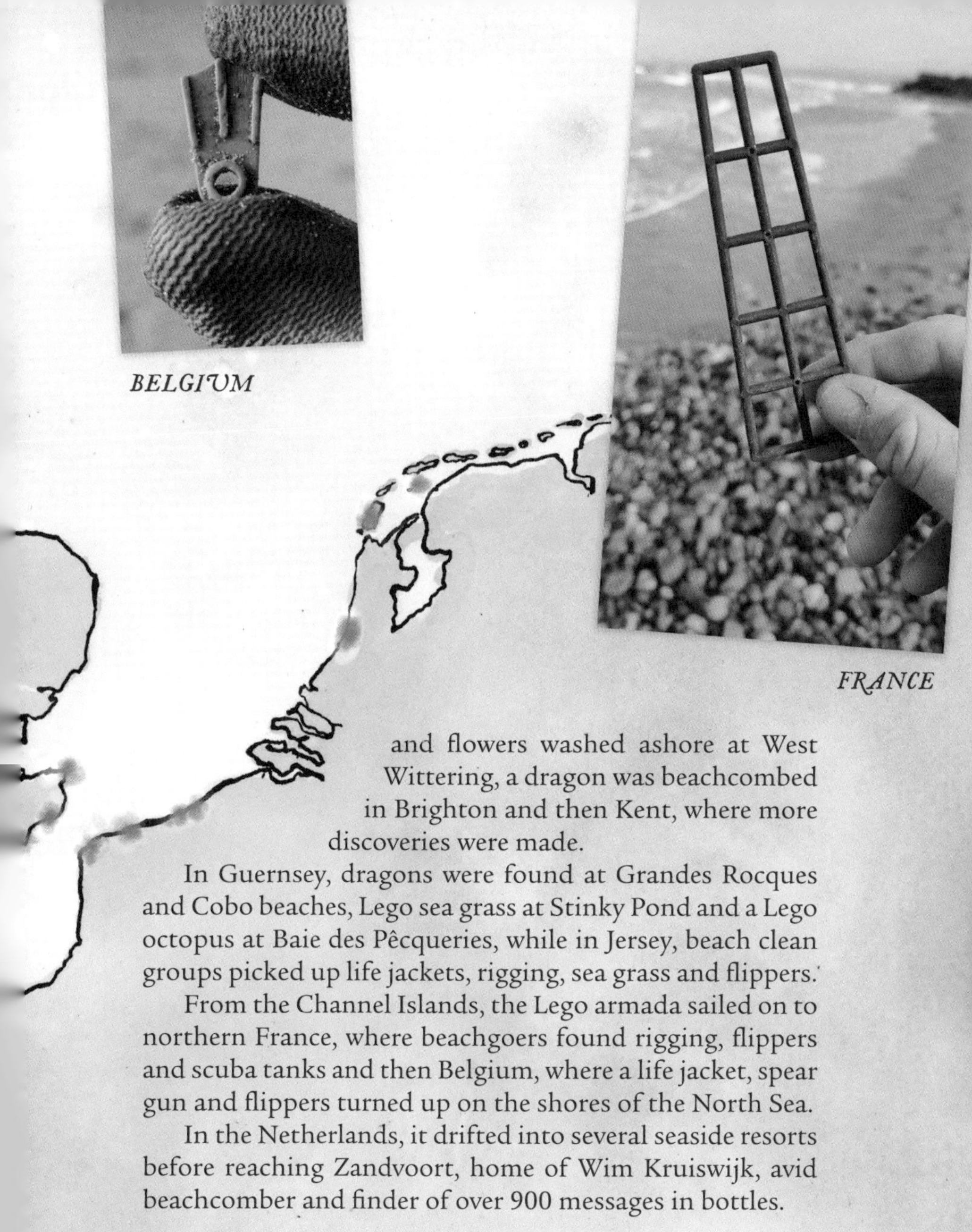
BELGIUM

FRANCE

and flowers washed ashore at West Wittering, a dragon was beachcombed in Brighton and then Kent, where more discoveries were made.

In Guernsey, dragons were found at Grandes Rocques and Cobo beaches, Lego sea grass at Stinky Pond and a Lego octopus at Baie des Pêcqueries, while in Jersey, beach clean groups picked up life jackets, rigging, sea grass and flippers.

From the Channel Islands, the Lego armada sailed on to northern France, where beachgoers found rigging, flippers and scuba tanks and then Belgium, where a life jacket, spear gun and flippers turned up on the shores of the North Sea.

In the Netherlands, it drifted into several seaside resorts before reaching Zandvoort, home of Wim Kruiswijk, avid beachcomber and finder of over 900 messages in bottles.

# *Typing with the tides*

Wim Kruiswijk has been finding Lego from the *Tokio Express* for decades. He first heard about the Lego spill back in 1997, later going to a toy shop to pick up some Lego catalogues so he knew exactly what to look out for.

The following year, he spotted three Lego spear guns washed up on the beach near his home in Zandvoort, a popular seaside resort west of Amsterdam, and began meticulously recording every piece in his journal of beachcombed finds.

In 1999, he discovered two black Lego dragons, along with flowers, divers' flippers and a scuba tank. A third dragon, missing its head, was given to him by a young girl who found it during a beach clean the following year on

Terschelling, one of the West Frisian islands in the northern Netherlands. More Lego turned up during a beach clean on Texel, a nearby island.

Later in the year, Wim went on a cycling holiday round the Isle of Wight in the UK, finding nine pieces, including his first Lego cutlass. After that, he discovered Lego almost every year.

By remarkable coincidence, finding Lego from the spill wasn't Wim's first encounter with the *Tokio Express*. Over the years, he has discovered more than 900 messages in bottles, replying to each individually on his old manual typewriter.

In 1986, he found a message in a bottle thrown overboard from the *Tokio Express*. On the back of a printed form from the ship, headed 'dangerous goods list', was a child's crayon drawing; the sun shining, a palm tree in the background and a stickman drawing of the sender himself, five-year-old Martin.

Wim discovered Martin's father had been the captain of the *Tokio Express*, and Martin and his two older sisters had been allowed to travel with him on the ship to Singapore, Hong Kong and Malaysia.

On their way home, Martin threw his message in a bottle into the North Sea. Some two months later, Wim found it washed up on his local beach.

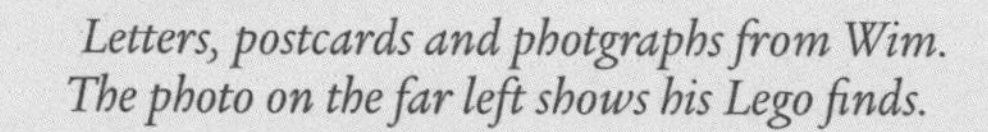

*Letters, postcards and photgraphs from Wim.*
*The photo on the far left shows his Lego finds.*

In 2015, a black Lego dragon was discovered on the small uninhabited Dutch island of Griend, in the Wadden Sea, by Job ten Horn from the Royal Netherlands Institute for Sea Research.

'We go to Griend during new moon spring tides to study shorebirds known as red knots,' he explained.

After finding the dragon, Job gave it to his girlfriend for her birthday. 'She's a researcher working on marine plastics ingested by northern fulmars,' he said. 'So it was not just a Lego dragon for her.'

*Job's Dragon*

Where the flotilla of Lego drifted next, we're not sure. Writing in his newsletter, *Beachcombers' Alert!*, back in 1997, Dr Ebbesmeyer predicted ocean currents would carry the Lego north past Norway into the Arctic Ocean, 'following the fabled Northeast Passage through the coastal waters of northern Siberia, arriving in Alaska after a dozen years'.

Some, he predicted, might even be carried by currents south to Japan and across the North Pacific Ocean to British Columbia, the westernmost province of Canada, and to Washington, Oregon and California on the west coast of the USA.

'By 2020, currents would have distributed Lego elements throughout much of the northern hemisphere,' he forecast.

Whether any of the Lego from the *Tokio Express* reached the Arctic we don't know, but polar expedition leader Dr Huw Lewis-Jones did once find a piece of Lego on a beach in Siberia. It wasn't from the spill, though. 'It was a tiny white brick,' he recalls. 'We also found a plastic hubcap and a bunch of beer cans.'

# CATCH OF THE DAY

Of all the stories that emerged after the BBC article, it was a fisherman's tale that intrigued the beachcombing community the most.

In September 2014, the partner of a trawlerman shared a picture of his 'catch of the day'. Spread out on the dashboard of a vehicle parked on a quayside was a haul of sea-worn Lego: shingled roof tiles, white door frames and bright yellow vehicle chassis we later discovered are from Lego dune buggies. This is Lego from the spill we haven't come across before; bricks that have been lying at the bottom of the sea since 1997.

As word goes round the harbour that we're on the lookout for Lego, the fishermen start saving it. 'If only we'd known earlier,' says one skipper, 'we could have given you sacks full of it.' It seems they have been finding it for years.

Soon I am making regular trips to Cornwall's ancient fishing villages, where I'm known by some as the 'Lego lady', meeting the trawlers as they come in. As well as picking up the Lego, I'm keen to find out what else has turned up in the nets, searching for clues as to what else may have been lost from the *Tokio Express*.

Over the whir of forklift trucks on the bustling quayside, the fishermen regale me with tales of Lego found long ago: dragons, octopuses, dragon wings and, it seems, packets of minifigure heads.

They show me what else they catch: ancient wine bottles, stoneware flagons and old terracotta jars, the shipping containers of their day. In the nets, too, are whalebones and one hundred-year-old fuel blocks once used on polar expeditions.

From the wheelhouses, the skippers send pictures of items hauled up but returned to the waves: wooden planks, tangles of cables resembling giant squid tentacles and huge doorframes from shipping containers.

One skipper describes the matching duvet sets he used to find, waterlogged but intact, that kept him in bed linen for years, while another speaks mysteriously of 'doing the pickle run'. Salvaged from the sea, too, are hose parts lost from the *Tokio Express* and sonobuoys – listening devices dropped by aircraft or ships carrying out anti-submarine warfare or underwater acoustic research.

And there are shoes, so many shoes. Once, when I visit, a skipper hands me a bucket full of waterlogged trainers hauled up from the ocean floor. The name on them is familiar: Ascot. Maybe they are from the *Cita*.

*Several of these Ascot trainers had hermit crabs inside when they were hauled up from the deep. The ocean dwellers had taken up residence after mistaking the shoes for shells. The crabs were later returned to the sea.*

Almost six weeks after the *Tokio Express* lost sixty-two shipping containers off Land's End, the bulk carrier *Cita* ran aground on the Isles of Scilly, just over 16 nautical miles away. The incident was said to have been caused in part by the ship's mate falling asleep while in charge of the vessel.

On board were 145 containers holding millions of pounds worth of cargo, including tobacco, tyres, golf bags, clothing, scales, Action Man clothes and accessories, textiles, trophy parts, barbecue sets, Ascot trainers, toilet seats and garden gnomes.

*Action Man clothing*

Although some of the containers lost overboard washed ashore, many were carried away by the tides, with fifty-five remaining unaccounted for, thought to have sunk

## *Fishing for Litter*

Many fishermen in the southwest of England and Scotland take part in Fishing For Litter, an award-winning project set up to reduce the amount of marine debris on beaches and in the ocean. The scheme is simple but effective: fishermen are provided with large, hardwearing bags to collect plastic waste found in their nets alongside the catch and bring it back to port. To date, over 1,866 tonnes of marine litter have been collected by boats taking part.

One of the more unusual finds reported by fishermen was the fibreglass body of an old Robin Reliant car. Too big to land, it was returned to the deep.

THE CONTAINER SHIP 'CITA' SINKS OFF ST MARY'S.

to the seabed. Among the goods that fell into the water were plastic carrier bags destined for an Irish supermarket chain, bearing the message 'Help protect the environment'.

At the time, there was confusion as to what cargo had come from the *Tokio Express* and what was from the *Cita*, as the illustration (above) by Cornish artist André Ellis from 1997 shows, with goods from both vessels all mixed up together at the bottom of the ocean. Even today, over two decades later, fishermen still find cargo from both ships jumbled up in their nets.

*Trophy parts*

# BREAKING NEWS
## *BENEATH THE WAVES*

As I turn on my computer one night, there's a message waiting for me. It's from a trawler fisherman. 'Found this chap last trip. He's 'armless, so won't bite.' In the accompanying picture is a sea-weary Lego man, missing an arm and his legs, the letters 'TV' on his chest.

I recognise him immediately. He's from the Crisis News Crew, a set released by Lego in 1997 containing a helicopter, a van, a pilot, a cameraman, a female news anchor and two cameras.

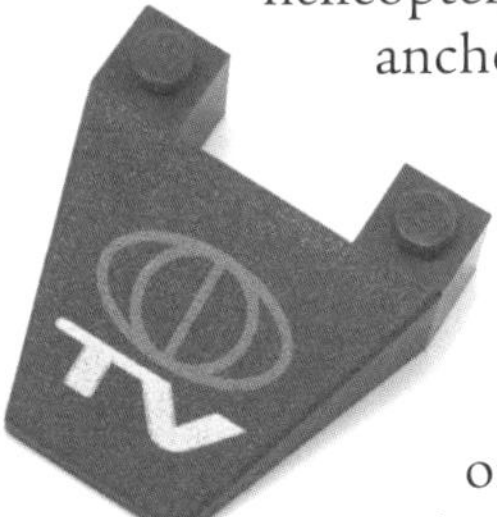

For a fleeting moment I wonder if they were filming as the container fell off the ship. Have they been reporting from the depths of the ocean ever since?

Of all the Lego figures that could have been on board the *Tokio Express*, it strikes me as a bizarre coincidence that – at a time when plastic pollution poses such a huge threat to the ocean – it was the crisis news crews who plummeted overboard.

Later I find out there were 59,500 Lego cameramen and pilots in the shipping container, but this is the only one I have ever seen.

*Quite a few of these 'blue slope bricks' have washed ashore. They're the bonnets for the Crisis News Crews' vans. There were 34,500 on board the* Tokio Express. *Although they don't float, this brick was found some 55 nautical miles from the site of the spill, along with other pieces from the shipping container.*

It wasn't just fishermen who were finding Lego we hadn't seen before, though. Beachcombers, too, had begun to discover pieces we didn't recognise. Not the familiar flippers, flowers and all the other Lego that had been turning up for years, but other pieces not featured on Dr Ebbesmeyer's earlier list of Lego bricks to look out for.

Mingled in with the seaweed were bright yellow portholes, parts of Lego submarines, white doorframes, blue panels and minifigure heads.

Then, in the spring of 2015, a marine debris artist picking up plastic from a harbour on the north coast of Cornwall discovered an odd assortment of sea-worn Lego.

Amid drifts of kelp, Michelle Costello spotted two tiny Lego storage containers, a log wall from a pirate fort, a large rock panel or mountain base and a two-stud brick with a radio and microphone printed on one side.

Later we found out these little radios were only produced in 1997, for the Lego Outback Airstrip set, a model of one of the little airstrips scattered around the Australian Outback.

Intrigued by all the different Lego now turning up, I asked Dr Ebbesmeyer if there was a complete inventory of everything in the shipping container; not just the list of 'floating' Lego published in his newsletter soon after the spill but one that included all the bricks thought to have sunk, presumably lost without trace.

In reply, he emailed me a scan of the original manifest Lego had provided in 1997 with the box of samples. It consists of three typed pages of A4, divided into two columns and showing one hundred items in total. On the left are Lego part numbers. On the right, the number of pieces lost.

These days, it's relatively easy to look up a bit of Lego on the internet. There are websites galore selling Lego bricks, many listing part numbers with a description and picture, but in the late 1990s linking a part number to a specific brick was far more difficult.

For Dr Ebbesmeyer, it was a case of studying the samples in the box and trying to match them with the part numbers on the list, writing in longhand alongside each his best guess at what each piece might be.

While it was obvious what some were, such as the dragons and octopuses, others were much harder to identify. To complicate matters, several paper labels had come off in transit and were now either loose in the box or stuck to each other.

It wasn't straightforward. The list of descriptions was incomplete and – without the box of samples – I too was struggling to fathom out what pieces the part numbers related to. It was the heading that immediately caught my eye though. Lego had titled the inventory:

LOST LEGO ELEMENTS
SITUATED: AT THE BOTTOM OF THE SEA.

# 3

# A Plastic Tide

## *All over our land and seas*

'If seven maids with seven mops
Swept it for half a year,
Do you suppose,' the Walrus said,
'That they could get it clear?'
'I doubt it,' said the Carpenter,
And shed a bitter tear.

extract from
The Walrus and the Carpenter
LEWIS CARROLL

The person who probably holds the record for finding the most Lego flippers from the spill is Rob Arnold, an environmental campaigner from Cornwall who uses the debris he picks up from beaches to draw attention to the problem of ocean plastic.

Over the years, I have accompanied Rob on some of his extreme beach cleans. I've been down on my hands and knees at dusk sweeping up plastic from the sand with a dustpan and brush, trailed anxiously behind him as he leapt across rocks to reach a remote beach strewn with plastic bottles and hauled myself up a steep, overgrown cliff path by rope, chased by an incoming tide.

Once, Rob and I ventured deep into a dark, lifeless sea cave where plastic debris had been accumulating for decades. Wearing hard hats, head torches and high-vis vests, we filled dumpy sacks with over 500 plastic bottles, some more than fifty years old.

The bottles were later collected by the Ocean Recovery Project* and used with plastic gathered from beach cleans, community litter picks and fishing ports to create a dance arena made from recycled plastic in the Shangri-La area at Glastonbury Festival.

Between 2017 and 2020, Rob and fellow beach cleaners picked up an estimated 25 million tiny bits of plastic from one beach in Cornwall, sweeping it all up by hand to make sure as much of the natural material as possible was left

**A programme launched by Keep Britain Tidy in the southwest of England to recover and recycle litter collected by volunteers on beach cleans.*

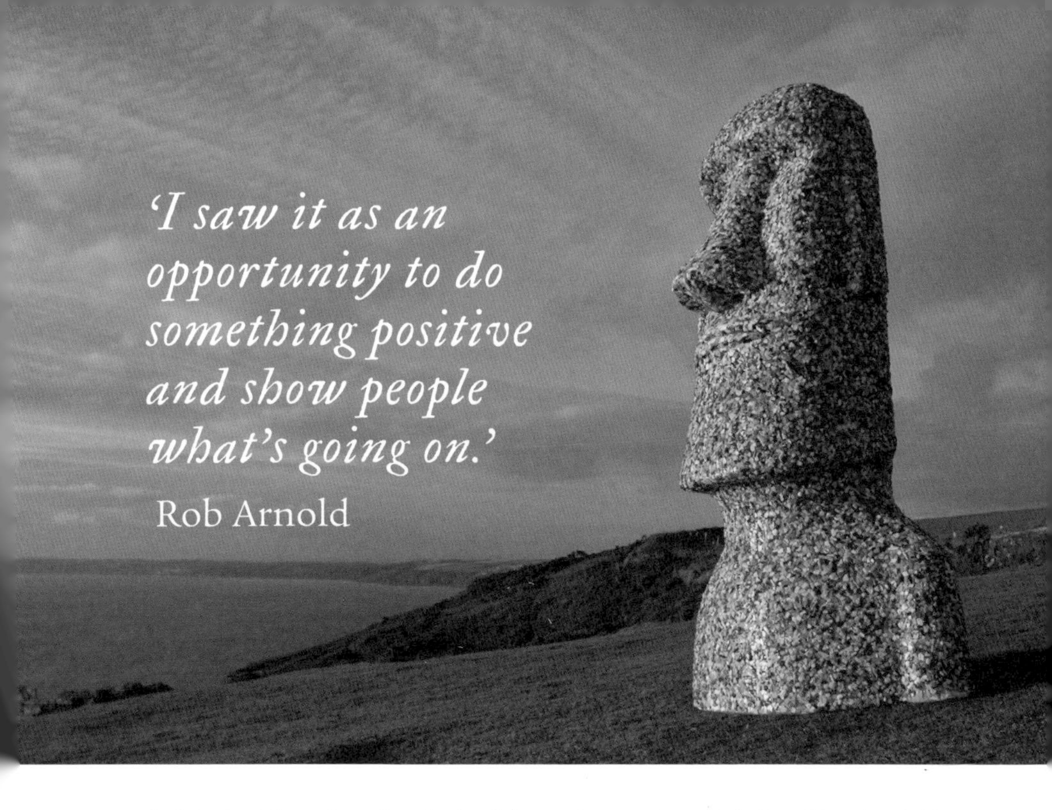

*'I saw it as an opportunity to do something positive and show people what's going on.'*
Rob Arnold

behind. After the plastic had been bagged up, Rob took it back to his workshop where he sieved it into different sizes to see how much he could identify.

Among the 120 sacks of plastic picked up were over 15 million nurdles and bio-beads, roughly 10 million fragments of broken plastic and polystyrene, toys dating back to the 1950s, almost 500 car tyre dust caps, a bucketful of cotton bud sticks and 1,430 bits of Lego, including nearly 1,000 flippers and 67 cutlasses from the spill.

*Rob Arnold sculpted this Moai Easter Island head replica from half a million pieces of plastic collected from Cornish beaches. Included are flippers, spear guns, flowers and cutlasses from the Lego spill. Rob sometimes takes the sculpture to marine conservation events, using it to call attention to the environmental impact of plastic in the ocean.*

This terrazzo-effect Gyro table was created by Tasmanian designer Brodie Neill from plastic salvaged from beaches in Hawaii, Tasmania and Cornwall. Nearly 250,000 tiny plastic fragments used in the table came from two beaches in Cornwall, swept up with a dustpan and brush by Rob Arnold, along with volunteers from Rame Peninsula Beach Care and the Lego Lost At Sea Project. Among the 9kg of plastic picked up from the sand were 23 pieces of Lego from the spill. Later, Rob separated out all the black, blue and white fragments and sent them to Brodie, who teamed up with scientists to turn all the plastic into a usable material. 'I wanted to draw attention to the potential of a material that is polluting the world's oceans.' said Brodie.

Artists who have featured Lego from the spill in their work include photographer and beachcomber Caroline South, who often uses hidden treasures picked up from the shores of West Sussex in her eye-catching colour-themed compositions to draw attention to ocean plastic.

Included in this arrangement are a Lego flower, flippers and sea grass, all found by Caroline while exploring beaches along the south coast of England with her family.

# Debris from the Deep

Overnight, strong winds have brought vast amounts of kelp ashore. It lies in drifts several feet deep down one side of the beach, banked up against the rocks, stranded by the ebbing tide.

Scattered throughout it is debris from the bottom of the sea. This is where the plastic that doesn't float ends up.

Old clothes, combs and the remains of mobile phones. Keys from typewriters, computer keyboards and calculators. Carpet remnants and golf balls. Cable ties, hairbands and frisbees. Flying discs and lost dog toys. Divers' fins, curtain hooks, deflated footballs. Bank cards, razor heads, medical lancets. Old baking powder submarines from cereal packets. And once, a witch's nose and a shrunken mask of Frankenstein's creature, the neck bolts and brow stitches still clearly visible.

As a child, I was frightened of wading through seaweed, feeling the fronds wrapping around my legs. It's known as fykiaphobia – the fear of seaweed – though these days I am more afraid of what lurks within it, rather than the seaweed itself.

There are fragments of plastic bottles too, thousands and thousands of them. While bottle caps often float ashore, the bottles themselves sink to the bottom of the sea, breaking apart into smaller and smaller pieces. Sometimes all we find are their faded bases, looking curiously like stranded jellyfish.

A friend joins me and we fill sacks with plastic. Later we count it. 1,097 plastic bottle fragments, 158 bits of broken fishing float, seventy-four sets of goggles, pairs of sunglasses and snorkelling

*Clockwise from top: witch's nose with warts and whiskers, flying discs, computer keys, disposable lancets*

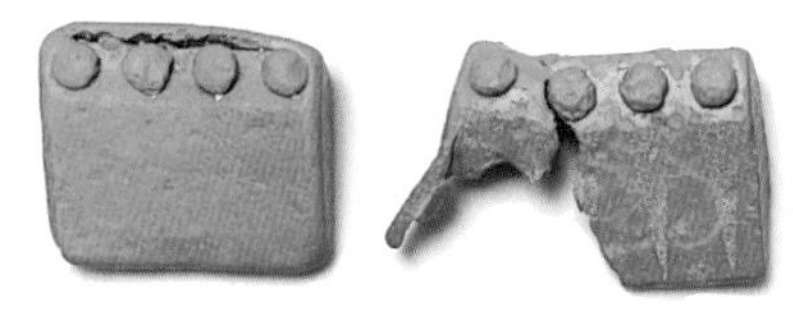  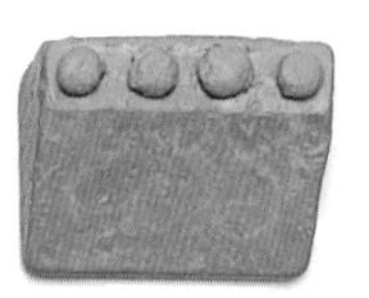

masks, thirty-two shoes and 211 strips of synthetic tyre rubber, perhaps from crab or lobster pots.

There is Lego from the *Tokio Express*, too. On this beach, about 55 nautical miles from the place where the containers toppled into the sea, these 'green slope bricks' often turn up. There were nearly 600,000 of them on board, some plain, others with yellow triangles or chevrons.

Whereas some of the Lego from the *Tokio Express* is in near perfect condition when found, these green bricks are warped and distorted after years drifting along the seabed.

## *sea fangles – tiny trees of the sea*

Among the kelp, beachcombers often find 'sea fangles', the name given to pink sea fans tangled in marine debris. A soft coral, pink sea fans are nationally protected, providing important habitat for fish such as wrasse, as well as nudibranchs or sea slugs and small-spotted catsharks, which attach their mermaids' purses or egg-cases to them.

After storms, pink sea fans wash ashore, their skeletons draped with everything from fishing nets, ropes and angling lines to tights, designer underpants, hoodies and woolly hats. Once, when a friend unravelled one, she discovered a piece of Lego sea grass from the spill among the strands of fishing rope snagged on it.

Not a tiny tree as I once thought but a colony of little animals. Because they are protected under the Wildlife and Countryside Act 1981, it is an offence to be in possession of – or sell – pink sea fans.

*Small-spotted catshark eggcase*

Found in one beach clean:
32 Shoes
1,097 Plastic bottle fragments
211 Strips of synthetic tyre rubber
74 Sets of sunglasses, snorkelling masks and goggles
158 Bits of broken net float

# THE ILL-FATED VOYAGE OF THE LITTLE LEGO LIFE RAFTS

It is among the kelp, too, in torrential rain in 2015, that I spot one of the fabled Lego life rafts. I can hardly believe it.

In over eighteen years of scouring the beaches, it's the only one I have ever seen. For years I have imagined them adrift in the ocean, voyaging to the Americas, ravaged by roaring seas, wrecked on rocks. But there it is, lying unceremoniously among the seaweed.

Then, a day later, on the same beach, in almost exactly the same place, a fellow beachcomber found another. And in the weeks that followed, two more were discovered a few miles along the coast. Where had they been all these years?

At first, I assume by a strange quirk of the tides they have all floated ashore together, ending an epic eighteen-year voyage in convoy. It's only later, when a fisherman sends a picture of one hauled up in his nets 20 miles offshore, that a thought occurs to me.

Maybe they never floated at all.

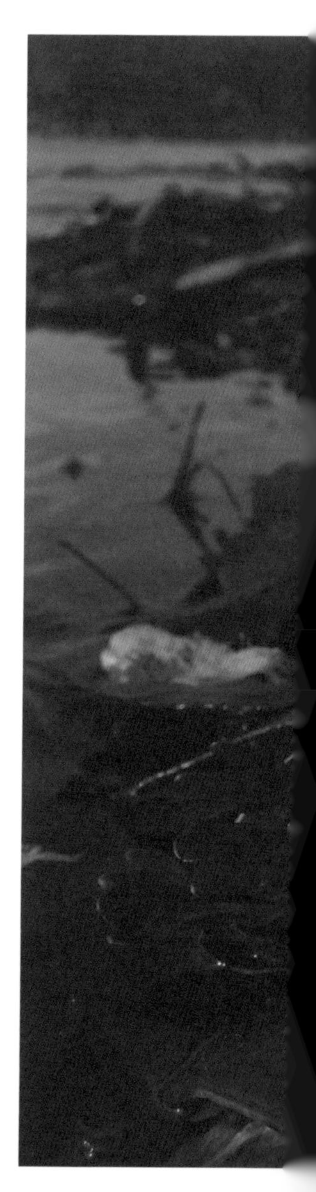

*There were 28,700 Lego life rafts in the container that fell off the* Tokio Express. *This one was found some 55 nautical miles from the site of the spill, eighteen years after it fell into the water. It no longer floats.*

# LIFE BENEATH THE WAVES

Man-made objects that sink to the bottom of the sea often act as artificial reefs, providing refuge for all manner of marine life. This Lego life raft, found in a fisherman's net, had been colonised by a cup coral, tube worms and bryozoans, tiny aquatic animals that form colonies.

◆ Said to be perfect for creating underwater landscapes, reefs and castle walls, these big ugly rock pieces and little ugly rock pieces are known as BURPs and LURPs by AFOLs (Adult Fans of Lego). They featured in several Lego sets from 1997, including Witch's Magic Manor, Deep Sea Bounty and Pirates Perilous Pitfall.

There were 18,720 on board the *Tokio Express*, 11,520 big ugly rock pieces and 7,200 little ones. These were found in fishermen's nets 20 miles off the Cornish coast in 2021.

◆ Used to create the reef in the Lego set Deep Reef Refuge, these little ugly rock pieces from the *Tokio Express* have become miniature reefs in their own right, colonised by marine creatures, including a saltwater clam, sea squirts and saddle oysters.

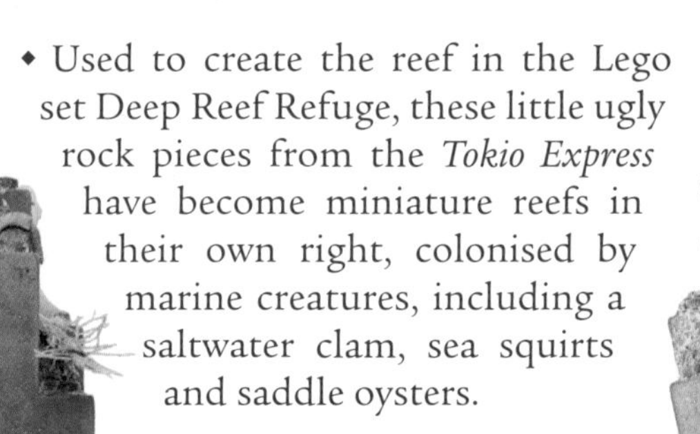

CHAD
WILD PACIFIC HALIBUT
CANADA
06 0727
H.L.
J104
RONSON
FLINTS
Ferguson

# *Gunpowder, Rum and Lego*

In 2018, divers from the Maritime Archaeology Sea Trust (MAST) and Bournemouth University made an unexpected discovery on the wreck site of HMS *Invincible*, the warship that sank in the Solent in 1758.

As well as a gunpowder barrel, swivel guns, a sandglass (hour glass) used to calculate the ship's speed, woodworking tools, a bottle of corked rum and a sailor's shoe, they also found two pieces of Lego.

'The wreck site was first excavated in the 1980s and 1990s,' explains maritime archaeologist Giles Richardson, the *Invincible* project diver who recovered the Lego from the sea floor. 'After the original excavations, the trenches were allowed to fill in naturally, so any rubbish rolling around the Solent, such as drinks cans, plastic waste and a few Lego bricks ended up sealed alongside the wreck for almost thirty years.'

When Giles first shared the picture of Lego bricks recovered from the wreck site of HMS *Invincible* on social media, I hadn't realised he too had found Lego from the *Tokio Express*. It was only later that he mentioned the hoard of Lego he and his family collected while on holiday in Cornwall in 1998.

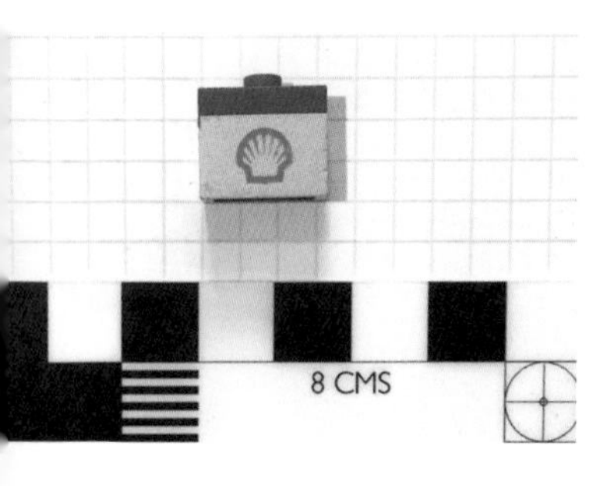


'We spent school holidays at a friend's home near Whitsand Bay,' recalls Giles. 'Exploring the beaches for treasure was a favourite pastime for years. It probably inspired my love of shipwrecks. The family Christmas newsletter that year proudly states we found 269 pieces of Lego. We all remember the spill being in the news, particularly people's excitement it was nautical themed pieces and disappointment it was only accessories but no actual Lego bricks, so you couldn't build a set from what you found.

'The dragons were of course what everyone wanted. When I went searching, some were easy to spot, just washed up on the sand or floating in rock pools. Others were wedged in rocks or caves after the winter storms and needed prying out of gaps. Even then some had been damaged beyond recognition.'

*A Lego dragon wedged between rocks – still there to this day*

The pieces of Lego Giles found on the wreck site of HMS *Invincible* are not from the *Tokio Express*, though. The brick with the Shell logo was first introduced in 1981.

# LEGO'S LIFESPAN IN THE SEA

Anyone who has ever trodden on a Lego brick will know how resilient they are – but how long could they last in the ocean?

In 2019, scientists from the University of Plymouth analysed fifty Lego bricks scooped up from Cornish coastlines to find out how long they could survive at sea. The pieces tested were not from the *Tokio Express* but older, classic style Lego bricks picked up during beach cleans.

In the lab, Dr Andrew Turner and his team washed, weighed and measured the bricks to see how much they

had weathered away, later analysing them to see what they were made from. As the recipe used by Lego for its bricks has changed over the years, working out the chemical formulation gives vital clues to their age.

After discovering many dated to the 1970s and 1980s, the scientists compared the bricks that had been in the sea for thirty to forty years to those from private collections from the same period that had not been underwater.

They found Lego bricks were so tough they could survive in the ocean for anywhere between one hundred and 1,300 years. 'Even we were surprised at how long Lego bricks lost at sea could last,' said Dr Turner.

## *How does Lego end up in the sea?*

Not all bricks found on beaches are from cargo spills. Some will have been left behind by children playing in the sand. Bricks dropped on streets could be washed into storm drains following heavy rain, gradually making their way into rivers and then the ocean. More could be from old landfill sites eroding into the sea.

However, a survey by Direct Line Home Insurance in 2016 suggested nearly 2.5 million Lego bricks could have been flushed down the loo by children, eventually ending up in waterways and oceans.

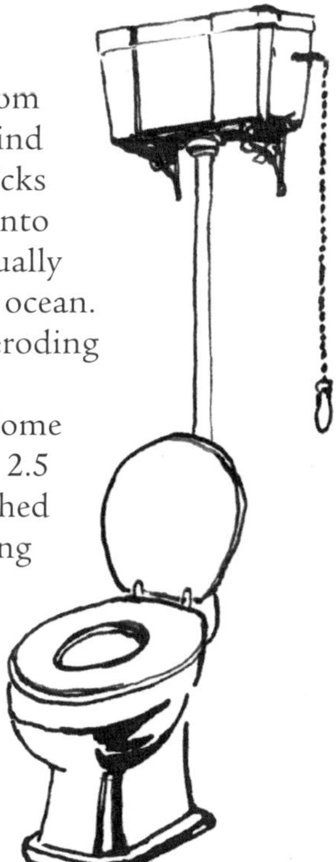

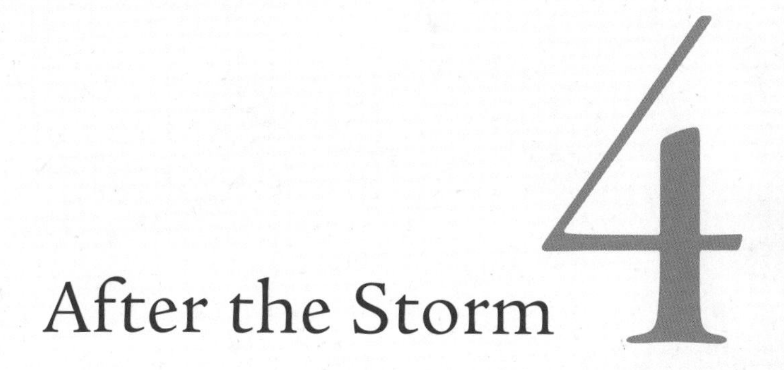

# After the Storm 4

## maggie and milly and molly and may

e.e. cummings

maggie and milly and molly and may
went down to the beach(to play one day)

and maggie discovered a shell that sang
so sweetly she couldn't remember her troubles,and

milly befriended a stranded star
whose rays five languid fingers were;

and molly was chased by a horrible thing
which raced sideways while blowing bubbles:and

may came home with a smooth round stone
as small as a world and as large as alone.

For whatever we lose(like a you or a me)
it's always ourselves we find in the sea

It's one of the highest tides of the year and the wind has been raging all night. I have been awake since the early hours, listening to the waves crash on the rocks below my house and the wind howling through the open windows, making the wooden rafters creak and lifting the tiles on the roof. Sometimes, when the blinds are billowing in the wind, it feels as though the house is setting sail.

Most days I scour the shores close to home, but this morning I am heading to a beach further along the coast. It was here in the sixth or seventh century that St Piran, an Irish priest and patron saint of tin miners, is said to have built his oratory in the dunes after floating across the sea from Ireland on a millstone.

It is still dark when I arrive. The wind is now gusting at 60 mph and my hair whips round my face, making it hard to see. I stumble down the long, winding track through the dunes, my dog racing ahead of me. The wind is rushing through the marram grass, leaving geometric patterns in the sand, barely visible in the moonlight. As I glance up, I can see the white crests of the waves in the blackness.

But I am too early. Huge waves are still surging forward, slicing through the dunes, leaving vertical walls of sand. It is too dangerous to drop down on to the shore, so I crouch among the spiky tufts of beach grass, waiting for the waves to recede.

Above me, a lone gull battles against the wind. Below me, debris washes in and out. Fishing floats, buoys and shoes are temporarily stranded and then washed back out to sea. A stray entrance funnel from a lobster pot rolls back and forth, carving parallel arcs into the sand before being reclaimed by the surging water.

At last, the tide begins to turn. I jump down on to the beach. There is plastic stretching for miles. Old plastic

bottles, knotted crisp packets, synthetic corks, yoghurt pots, paintbrushes, shotgun cartridges, toothbrushes, hairbrushes, hair curlers, foam strips from commercial longline fishing for swordfish and tuna, glowsticks, torches and coat hangers.

A sea of tiny polystyrene fragments stretches out in front of me. It's a menace on beaches and virtually impossible to remove. It comes from cheap styrene foam bodyboards that often break the first time they are used: from fishing floats, fish boxes, floating pontoons and from refrigerated containers lost to the waves. It gets caught in the seaweed, blown through the dunes and clings to the surfaces of caves, twinkling like stars at night.

On days like this it's impossible to pick up everything, so I focus on lighters (253), plastic bottles and containers (236), fluorescent light starters (40), plastic corks (45), shoes (27) and hundreds of blue stoppers, thought to be from ale casks. To date I've picked up over 20,000.

Among the swathes of plastic are older artefacts too: an old glass Toshiba headlamp and ancient wooden deadeyes once used in the rigging of a traditional sailing ship, so called because the three holes resemble the eye and nose sockets of a human skull.

As usual I have bags for debris and bags for treasure. In the 'finds' bag is a toy soldier that could date back to the 1950s, cereal packet toys from the 1960s, Lyon's Maid Superheroes lolly

*A Duplo figure among polystyrene fragments*

sticks from the 1970s, and over seventy Smartie lids, some nearly sixty years old. There is Lego from the spill too. Sea grass, spear guns, flowers, brooms, life jackets and rigging from the Witch's Windship, over seventy pieces in all.

The best find comes on the way back to the car, however. Among the mounds of plastic and polystyrene strewn across the sand I come across the greatest prize of all – that most precious of drift seeds – a Mary's bean.

Nickar nuts

# Sea Beans

*Magical gifts from the sea*

Sea beans or drift seeds travel thousands of miles, carried to remote shores by ocean currents.

Manicaria saccifera
*Golf ball bean*

Said to bring good luck to those who find them, these exotic objects are the fruits and seeds of tropical plants such as climbing vines, shrubs and palm trees. Many begin their journey in the rainforests of Central and South America, where they fall into swamps and rivers before being swept out to sea.

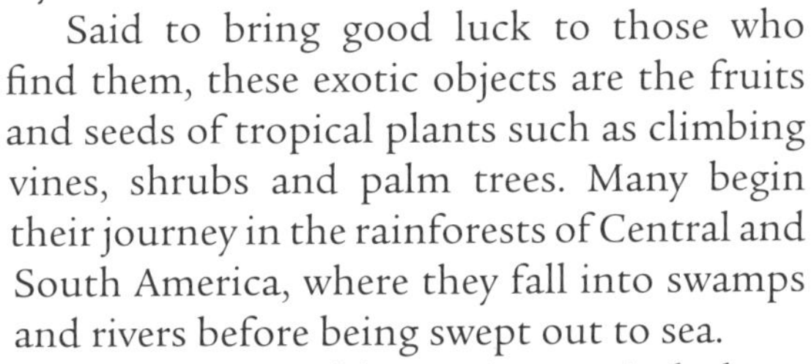

There's something quite magical about finding a sea bean. To hold it in your hand, press it against your cheek and imagine how far it has drifted. Beautiful sea hearts, glossy hamburger beans and the rare and highly prized Mary's bean, polished like ebony.

There's a drawing of a Mary's bean in an ancient tome (opposite) in Orkney Library and Archive. It's in *Ane Account off the Ancient and Present State off Orkney*, handwritten and illustrated by Mr James Wallace, Minister of Kirkwall, in 1684.

In Chapter 2, headed 'of Plants animals metalls & Substances cast up by the sea etc', Wallace writes: 'Sometyme they find living tortoises on the shoare and sometyme Tangles full of shells, evrie one haveing a pearle in them & verie oft these prettie Nutts off which they

*Hamburger beans*

*Sea heart*

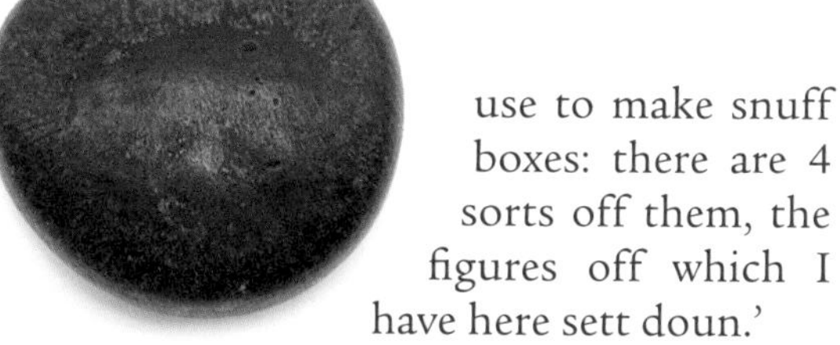

use to make snuff boxes: there are 4 sorts off them, the figures off which I have here sett doun.'

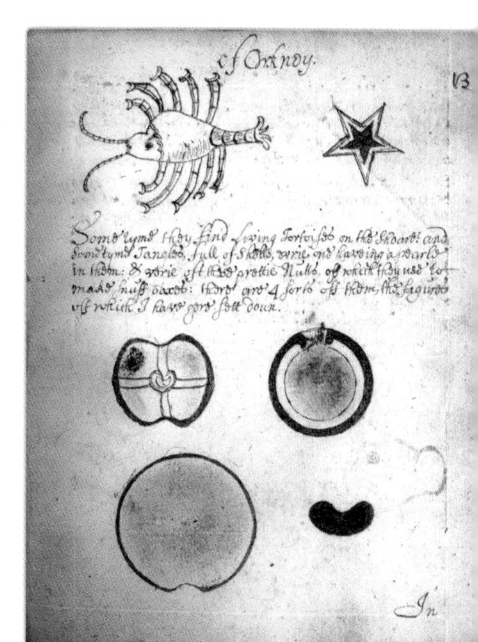

of Orkney.

13

Some tyms they find living Tortoises on the Shoare: and sometyms Tangles, full of Shells, verie and having a Pearle in them: & verie oft these prettie Nuts, of which they use to make snuff boxes: there are 4 sorts off them, the figures off which I have here sett doun.

Alongside, Wallace has drawn various drift seeds, including a Mary's bean. It's the first record of one washing ashore in the British Isles. Known, too, as crucifixion beans, because of the indented cross on one side, Mary's beans are steeped in myth and legend.

It is said a mother's childbirth pains will be soothed by clenching a Mary's bean in her hand.

Of all the drift seeds, the Mary's bean holds the record for travelling the farthest, carried by ocean currents from the rainforests of Mexico and Central America to Norway, a journey of 15,000 miles.

Over in Ireland, marine biologist Declan Quigley has been keeping a record of all Mary's beans found washed ashore in northwest Europe.

First on the list is the Reverend James Wallace's Mary's bean from about 1684. Next is one found in Orkney around 1696 by George Garden of Aberdeen and recorded by physician and naturalist Sir Hans Sloane, whose collection of objects from around the world became the founding collection of the British Museum.

In twenty-first place is one I discovered on the north coast of Cornwall in 2015, after initially mistaking it for a conker.

*Mary's beans*

# PLASTIC ADRIFT

Just as sea beans drift for thousands of miles on the surface of the ocean, so does plastic, carried to the shores of Cornwall and beyond by the Gulf Stream and North Atlantic Drift, currents that run like rivers through the sea.

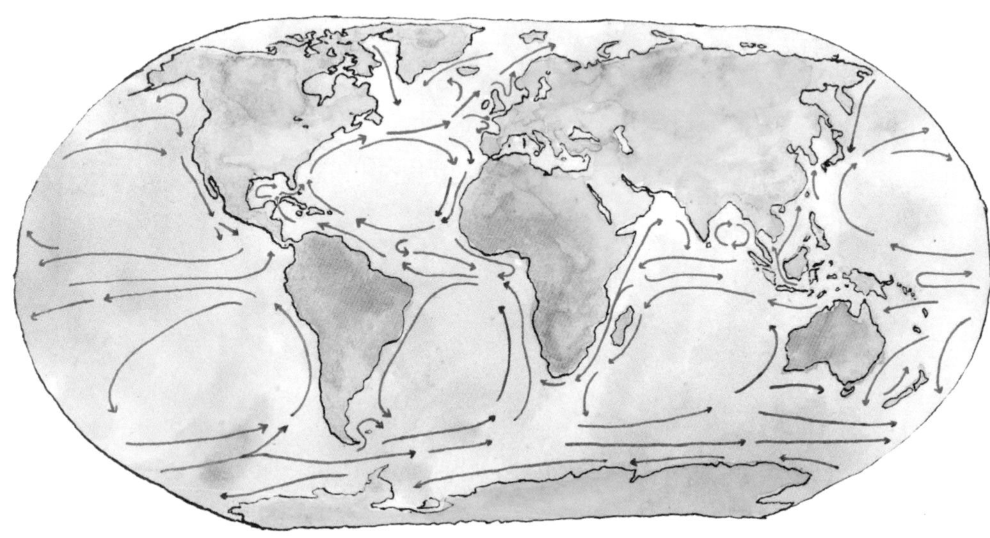

*Key: ● warm current ● cold current*

Fishing floats and buoys from the USA and Canada, lobster trap tags from Maine and Massachusetts, Nova Scotia and Newfoundland. Sometimes there are octopus pots too, dented and sea-worn. Once these would have been made of terracotta or earthenware. Now many are made of plastic, drifting vast distances from octopus

*Octopus pot*

fisheries in Spain, Portugal, Mauritania and Morocco after breaking free from the slab of cement inside that once served as ballast. Some have initials branded into them, providing clues as to their origin.

Occasionally, drifting plastic brings with it exotic species. In 2015, shells that a friend and I discovered inside a plastic bait jar washed ashore were identified as Stocky ceriths, a tropical sea snail not before recorded in the UK. Known, too, as Lettered horn shells, these are generally found among the shallow waters and reefs of South Carolina, southeast Florida, Bermuda, the West Indies and Brazil. It's thought they would have entered the bait jar as youngsters, feeding on algae and detritus. As they developed, they would have become trapped inside.

This Florida rock snail (below) was found alive, well and thousands of miles from its home in the Americas after hitching a ride across the Atlantic on a polystyrene fishing float. We later rehomed it at a local aquarium. A rare find for the UK and a new record for Cornwall.

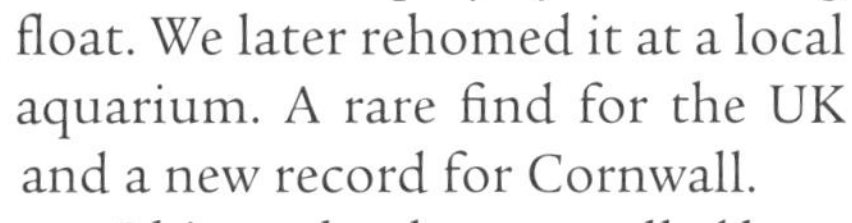

Objects that have travelled long distances in the open ocean often have goose barnacles attached, strange looking creatures with feather-like appendages or legs that they use to sweep the water for food. Their larvae drift in the sea, waiting to grasp on to something solid. In medieval times, it was thought goose barnacles

*'Always, then, in this flotsam and jetsam of the tide lines, we are reminded that a strange and different world lies offshore.'*

From *The Edge of the Sea* by Rachel Carson

*Goose barnacles attached to a sandal*

grew on trees and that barnacle geese emerged from the shells.

In days gone by, goose barnacles would have latched on to logs and seaweed. Nowadays, these ocean hitchhikers attach themselves to anything and everything, from toothbrushes and taps to bottles and buoys, their size often indicating how long they have been at sea.

Making discoveries from objects washed ashore isn't new. Legend has it the explorer Christopher Columbus was inspired to look for lands to the west in the fifteenth century after finding an exotic sea heart washed up on Europe's Atlantic coast. Today the sea heart is still known in the Azores as 'fava de Colom' or Columbus bean.

*This wooden drifter was found on the Holy Island of Lindisfarne off the northeast coast of England in September 2019. It is one of thousands released by the University of Oldenburg into the North Sea to map the movement of marine litter. Each has a message in German and English, together with a unique number and a website address where the find can be reported. The university hopes the research will help it understand current and future distribution patterns of plastic waste in the seas and where possible tackle its source.*

MICROCHIPPED
WITH TRACER
ANIMAL CODER
L·CY9
NISSAN

TV
TV

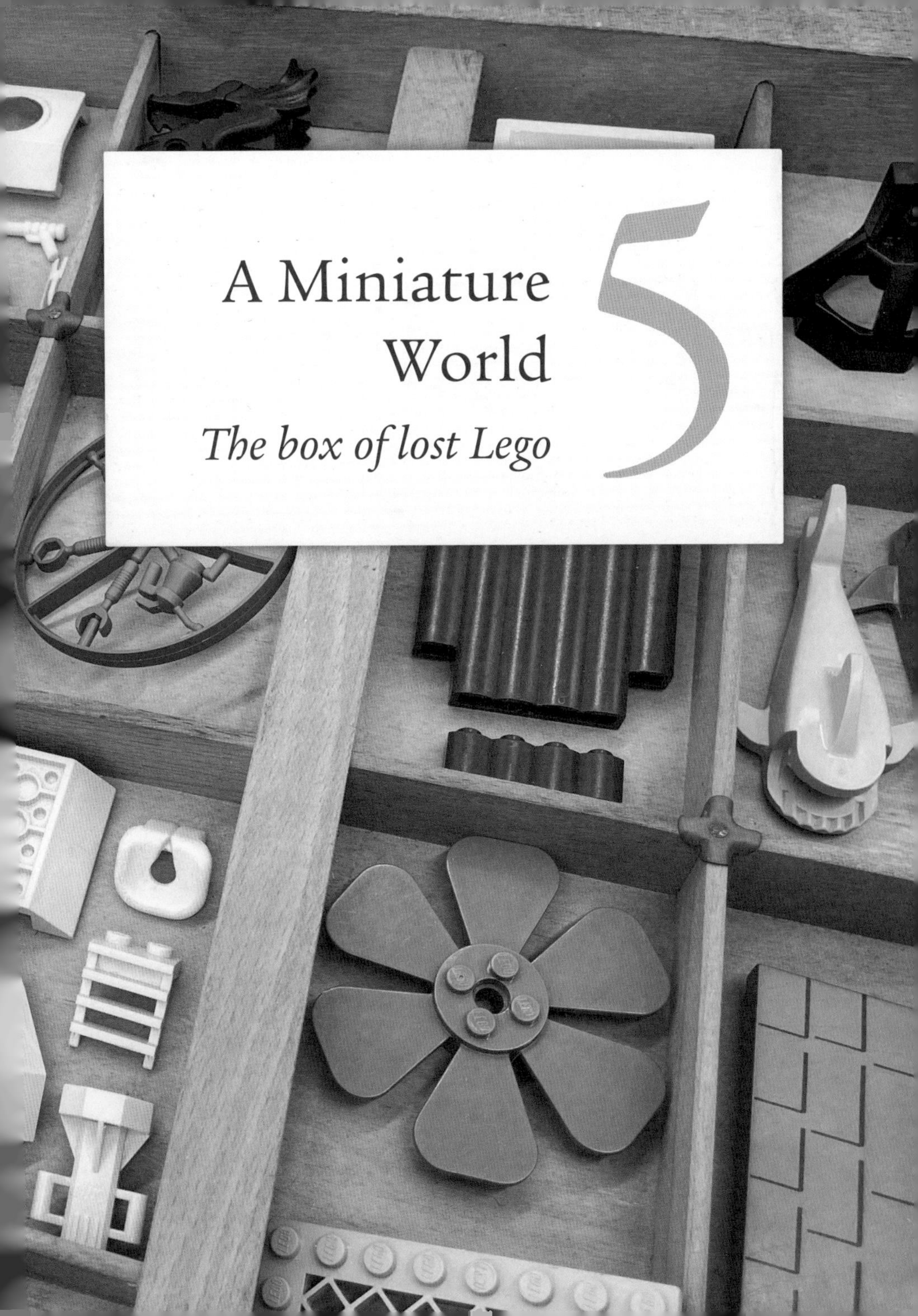

# 5 A Miniature World

## *The box of lost Lego*

'I'd send you the samples, but I'm worried about them getting lost,' says Dr Ebbesmeyer in an email. However, a year later, he generously ships them from the US after all, which is how a large cardboard box came to arrive for me in England in the summer of 2016.

The postman delivers it to the family home on the south coast of Devon, the old house on the cliff where my parents once lived. Now home to my stepmother, it's still the hub of family life, a place that anchors us all.

The bay below is full of shipwrecks. At night, you can see the beam of light from the Eddystone lighthouse sweeping across the horizon, warning mariners of the treacherous reef below. I've spent many an evening here gazing out to sea, imagining what lies beneath the waves.

In a bedroom upstairs hangs an old parchment map of wrecks lying off the South Devon coast. There's the *San Pedro El Mayor*, an Armada hospital ship carrying over 200 sick and wounded men, stranded in 1588; *The Dragon*, a British merchantman that sank in 1757; and the steamship *Jebba*, which went down in 1907 in thick fog on its way back from Sierra Leone. Thankfully, all crew and passengers on board were rescued, along with the ship's cat and two chimpanzees, carried ashore in blankets.

Over the years, divers have made amazing discoveries in the bay and beyond: ingots, cannons, ship rigging, anchors, portholes, guns and gold coins. Relics from a different time and another world.

The day I go to collect the box is grey and dull, a sea mist rolling in over the cliff. Waiting for me inside the house is a big white cardboard box bearing the words 'United States Postal Service'. According to the Customs Declaration and Dispatch note accompanying it, the box holds one hundred Lego pieces, total value $10.

Opening it is like delving into a long-lost treasure chest. There's even a miniature pirate's sea chest inside, along with four tiny gold coins and a diver, although whether these are from the spill isn't clear. It seems they may have been added later.

Within the box are all the different bricks Lego's shipping department sent Dr Ebbesmeyer in 1997, exact replicas of the pieces that tumbled into the sea during the storm. They're divided into two plastic storage containers. On top of one is written 'FLOATABLE LEGOS'. On the other, 'SINKER LEGO FROM SPILL'.

Some of the pieces are familiar. The black dragon, the flippers, the flowers, the rigging from the Witch's Windship, the life jacket, the cutlass and all the other bits of Lego beachcombers have been finding for years.

But among them are pieces I've never seen before. A shield and bat-winged helmet, a witch's hat and a forestman's cap. A cowboy's hat, a magic wand, a black bat. A pistol and a rifle.

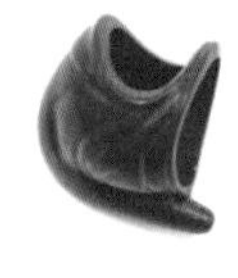

It's the sea-themed pieces that intrigue me most, however. A porthole

*Underwater scooter*

and a cannon base. Boat propellers, submarine parts and an underwater scooter. Bricks for creating reefs, rocks and underwater landscapes, log walls from pirate forts and, of course, the little yellow life raft.

A whole world of shipwrecks and seafaring adventures, in miniature.

I'm particularly taken by the tiny set of tools: a drill, an oil can, a hammer, a screwdriver and several wrenches. I can't help thinking it's just the sort of toolkit that might come in handy if you happened to be a little Lego man in a watery predicament, trapped in a rusting shipping container at the bottom of the sea.

Not all the pieces are there though. The green dragon has escaped and there's no sign of the octopus. Perhaps it squeezed its way out through the lid of the box.

Intrigued as I am, however, to be unravelling more of this Lego ocean mystery, my heart sinks a little as I realise just how much plastic was lost, how small some of the pieces are and what impact they could have on the environment and on marine life, which will not know the difference between a witch's hat and a worm.

# *The Lego that Fell into the Sea*

Among the millions of bits of Lego that sank into the abyss were black bats and witches' hats, frying pans and farmers' caps, dragon wings and magic wands. Shown here are all the different pieces listed on the inventory that Lego provided back in 1997, along with quantities. Many have never been seen since.

35,100

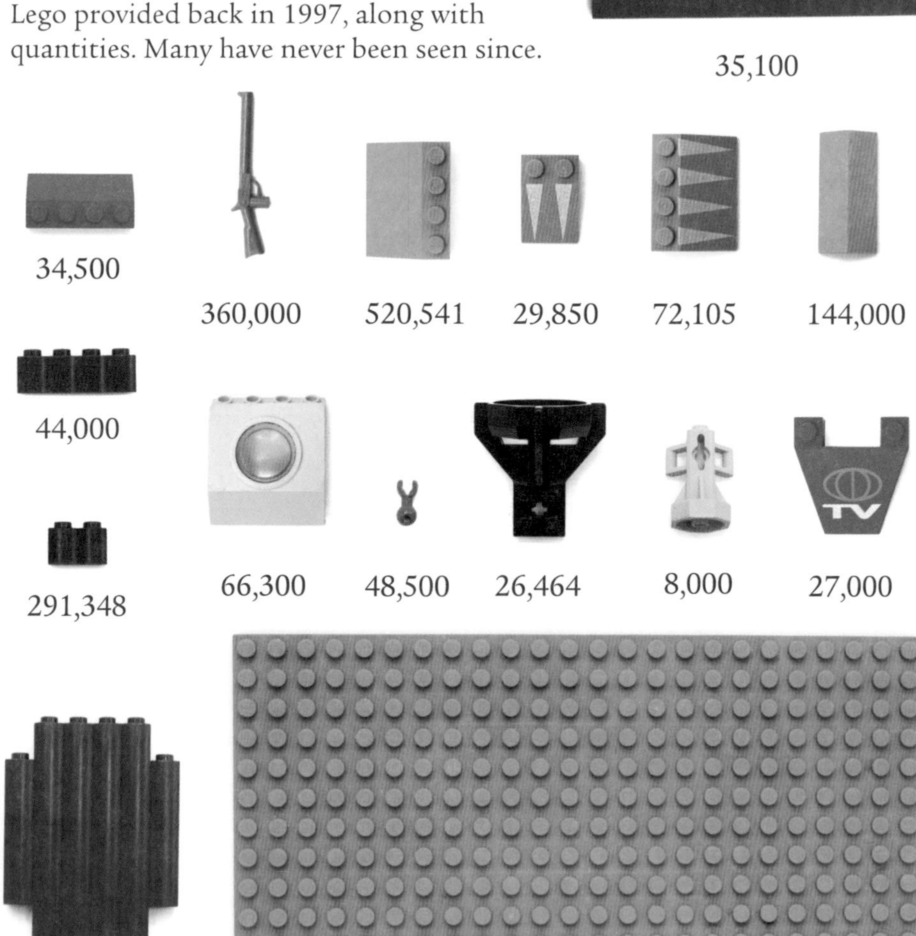


34,040

4,200

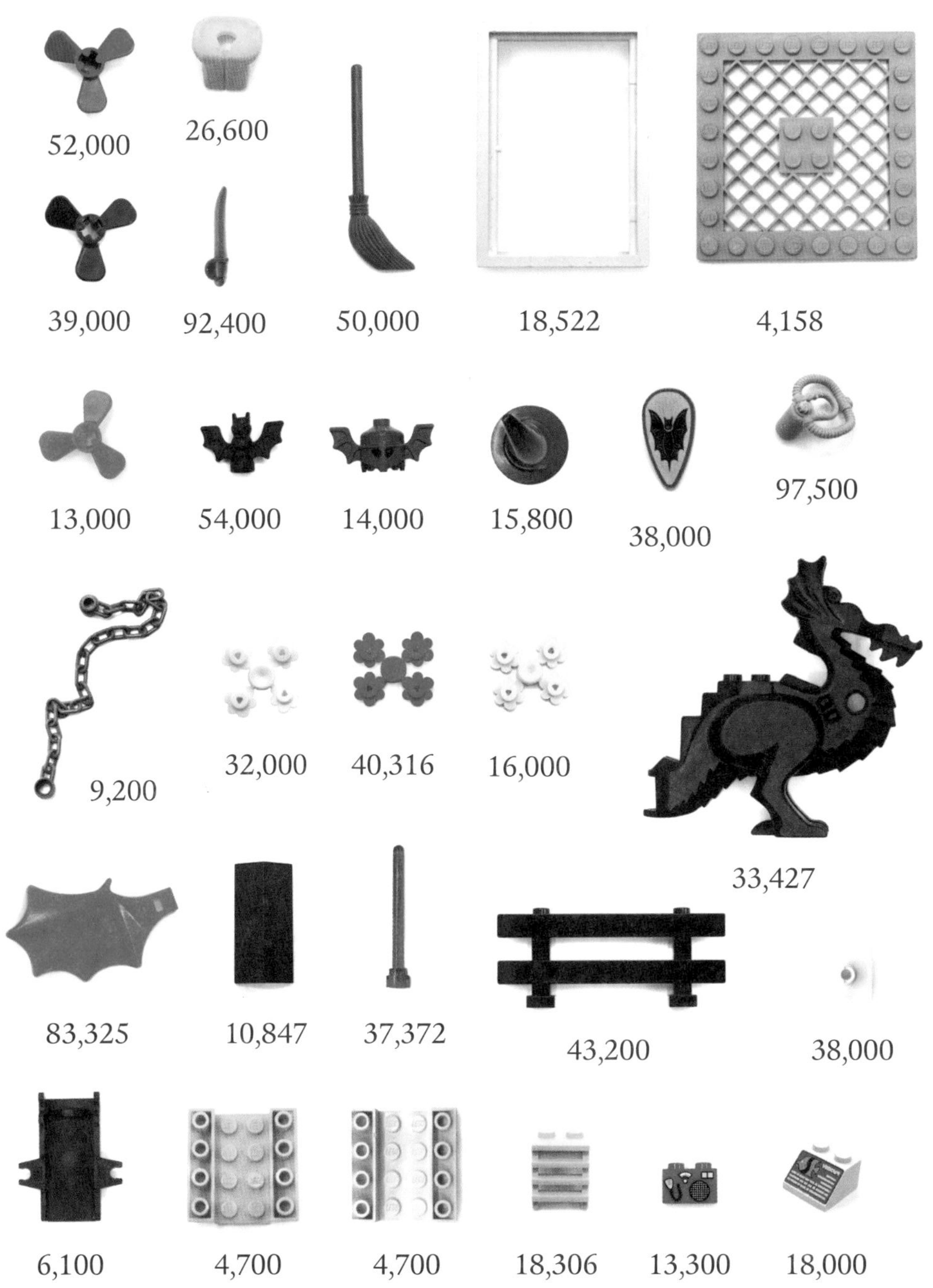
52,000
26,600
39,000
92,400
50,000
18,522
4,158
13,000
54,000
14,000
15,800
38,000
97,500
9,200
32,000
40,316
16,000
33,427
83,325
10,847
37,372
43,200
38,000
6,100
4,700
4,700
18,306
13,300
18,000

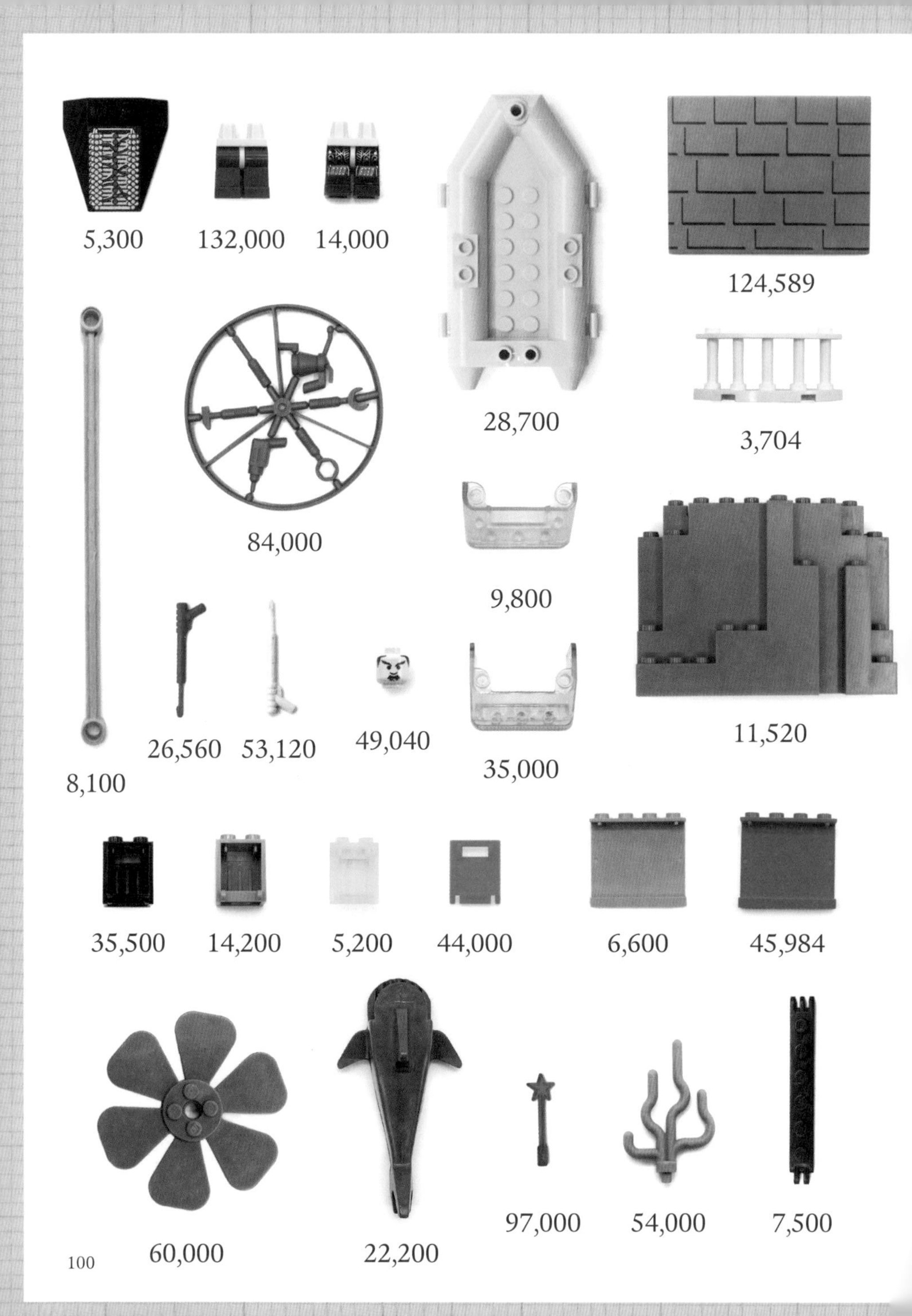
5,300
132,000
14,000
28,700
124,589
3,704
84,000
9,800
11,520
26,560
53,120
49,040
35,000
8,100
35,500
14,200
5,200
44,000
6,600
45,984
97,000
54,000
7,500
60,000
22,200

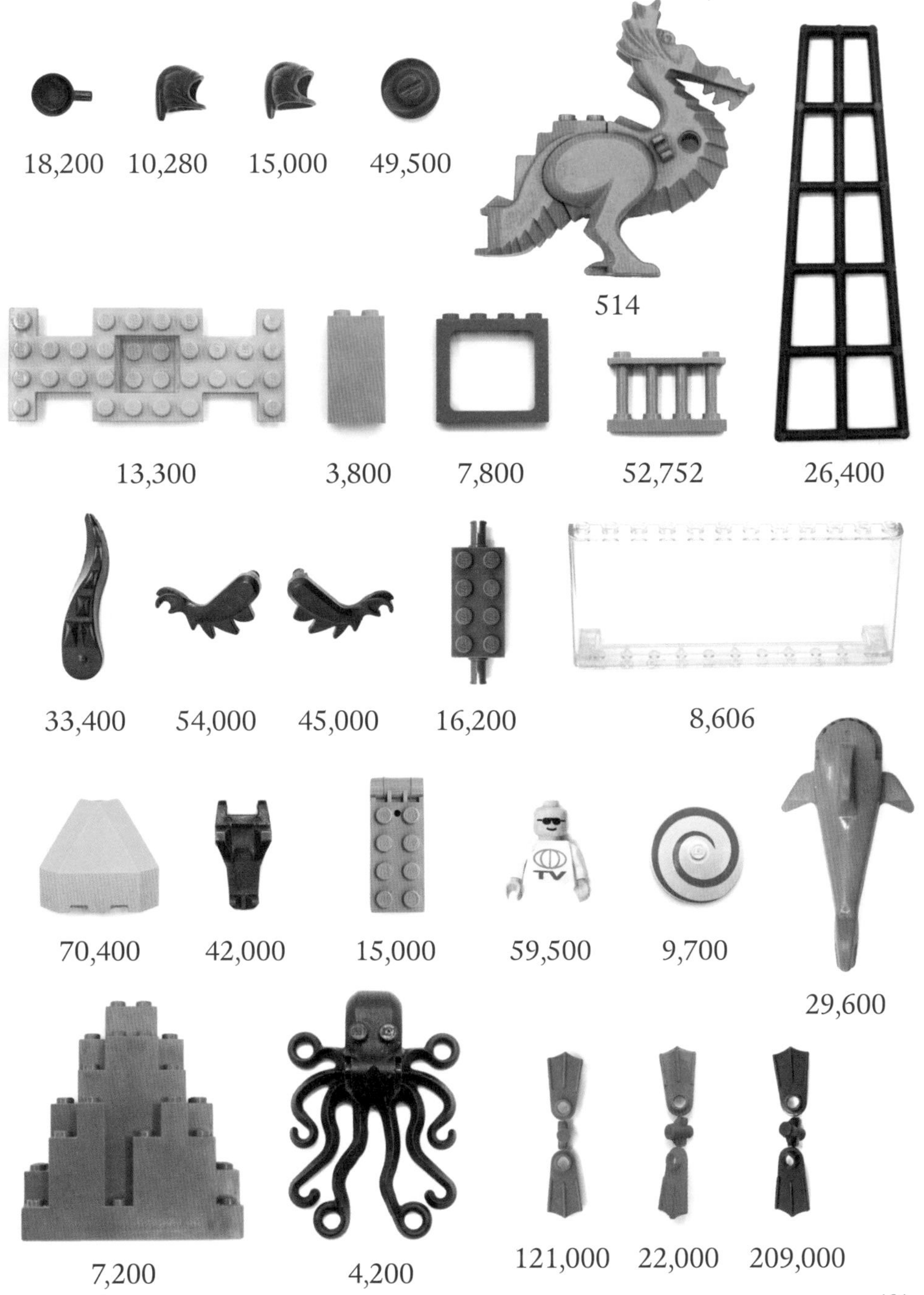
18,200
10,280
15,000
49,500
514
13,300
3,800
7,800
52,752
26,400
33,400
54,000
45,000
16,200
8,606
70,400
42,000
15,000
59,500
9,700
29,600
7,200
4,200
121,000
22,000
209,000

# WELL, SHIVER ME TIMBERS

It seems octopuses and dragons weren't the only creatures lost overboard from the *Tokio Express*.

Inside the container of Lego that plummeted into the sea were shivers of sharks – 22,200 dark grey and 29,600 light grey, 51,800 in total. Where are they all? None of the beachcombers or fishermen I have spoken to has ever found one.

Over coffee in a seafront café, I met up with a university lecturer who studies the long-distance journeys of animals and fish, including globally endangered basking sharks. During the conversation, the Lego spill comes up and I mention the sharks lost in the storm.

She looks thoughtful. 'I wonder if one day there will be more Lego sharks in the sea than real ones,' she says.

According to research published in the journal *Current Biology*, more than a third of all shark and ray species are now facing extinction.

*'Over a third of all shark and ray species now face extinction.'*

CURRENT BIOLOGY JOURNAL

FEED

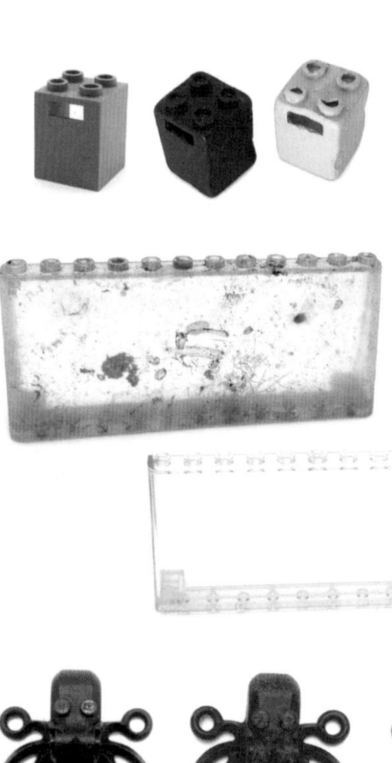

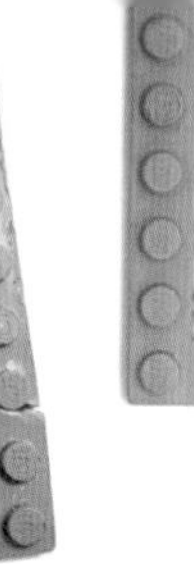
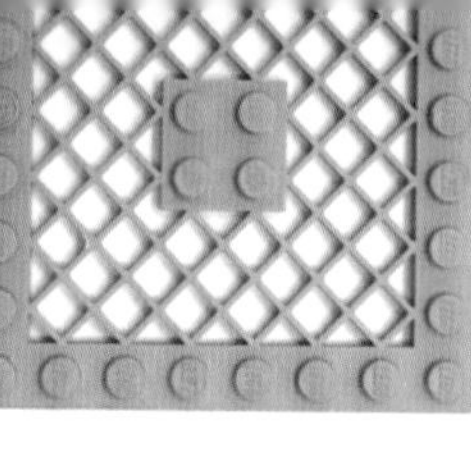

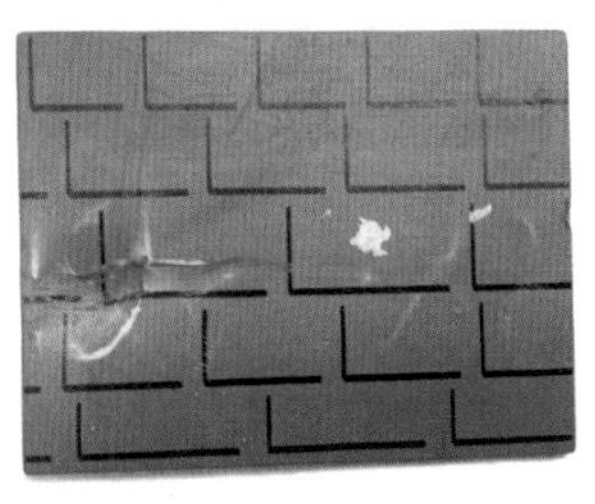
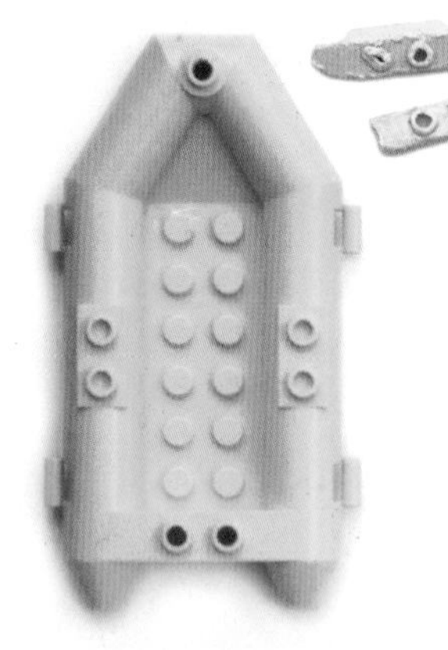

# How Lego has fared in the sand and sea

In the ocean, as on land, plastic breaks apart into smaller and smaller fragments. These pictures show what has happened to Lego from the spill over time.

Plastic is everywhere. It's in the air we breathe, in Arctic snow and Antarctic ice. It's in tap water, shellfish and beer. It's even been found in human placenta. It's in animals and it's in us.

In 2018, scientists spotted a plastic bag at the bottom of the Mariana Trench, the deepest part of the ocean. In 2020, tiny particles of plastic were found in snow close to the peak of Mount Everest, the world's highest mountain.

Every year between 8 and 13 million tonnes of plastic* enter the ocean. In the sea, plastic can drift for thousands of miles, caught up in ocean currents. Exposed to the sun's radiation and ocean waves, it breaks apart, becoming microplastics, tiny fragments less than 5mm in diameter.

However, only a small fraction of plastic that enters the ocean is said to be floating on the surface. The rest is

*There were 18,522 white Lego doorframes on board the* Tokio Express, *though few have been found by beachcombers. These were all found in fishermen's nets, hauled up from the bottom of the sea. They're thought to be the frames for the doors to the exploding bank vault in the Lego wild west set, Gold City Junction.*

thought to be still sinking, caught in the water column or lying at the bottom of the sea.

Every year, hundreds of thousands of marine creatures and seabirds die as a result of plastic pollution. They eat it, they become entangled in it. Seals, turtles, dolphins and porpoises have all died after becoming caught in fishing gear. A dead whale that washed up in the Philippines had 88lbs of plastic in its stomach.

Whether Lego from the spill has ever been found in the guts of marine creatures, I don't know, although fish have become entangled in it.

In 2018, an image emerged on social media of a small shark trapped inside a white Lego doorframe. The shark, a lesser-spotted dogfish or small-spotted catshark, had been discovered by a Cornish shell fisherman.

'Plastic pollution in real life at sea,' he wrote. 'Living a normal healthy life. A dogfish off Cornwall is caught with a Lego piece around it.'

Then another image of a dogfish trapped in a Lego doorframe surfaced. Had they swum through the frames while gliding along the ocean floor, scavenging for food? Or were they caught up together as the nets skimmed the ocean bed, only becoming entangled afterwards? As dogfish have rough, sandpaper like skin, it's easy to see how the frames could have become caught fast.

**Source: Marine Conservation Society.*

Great Bay of Eskimaux
C. Bluff
C Charles
STR. of BELL ISLE
Cape de Grat
White B.
Pinguins I
Anticosti
C. St. John
Cork B
GULF OF
Gaspee B.
NEW FOUNDLAND
C. Bonavista
Conception Bay
C. St. Francis
St. Johns
Magdalen Is
Is. St. Paul
C. Ray
The Great
The Flemish Cap
Track from England to New
a Breaker
Jaquet I. Doubtful
CAPE BRETON
Louisburg
C. Canso
Hallifax
Sable I.
belonging to the
Flores
Rock
a Bank of Breakers or
taken for Bermudas
Tropic of Cancer

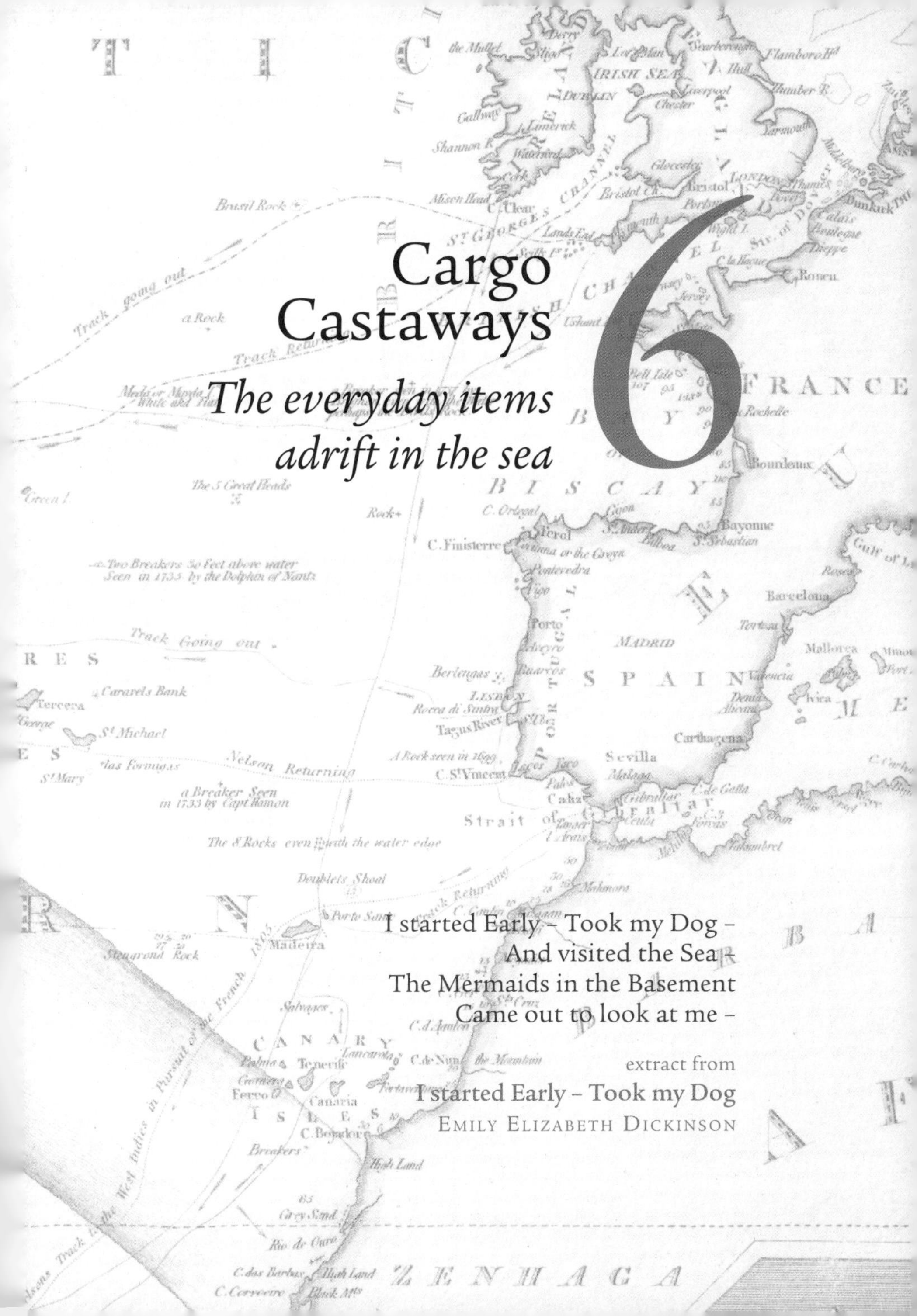


# 6

# Cargo Castaways

## *The everyday items adrift in the sea*

I started Early – Took my Dog –
And visited the Sea –
The Mermaids in the Basement
Came out to look at me –

extract from
I started Early – Took my Dog
Emily Elizabeth Dickinson

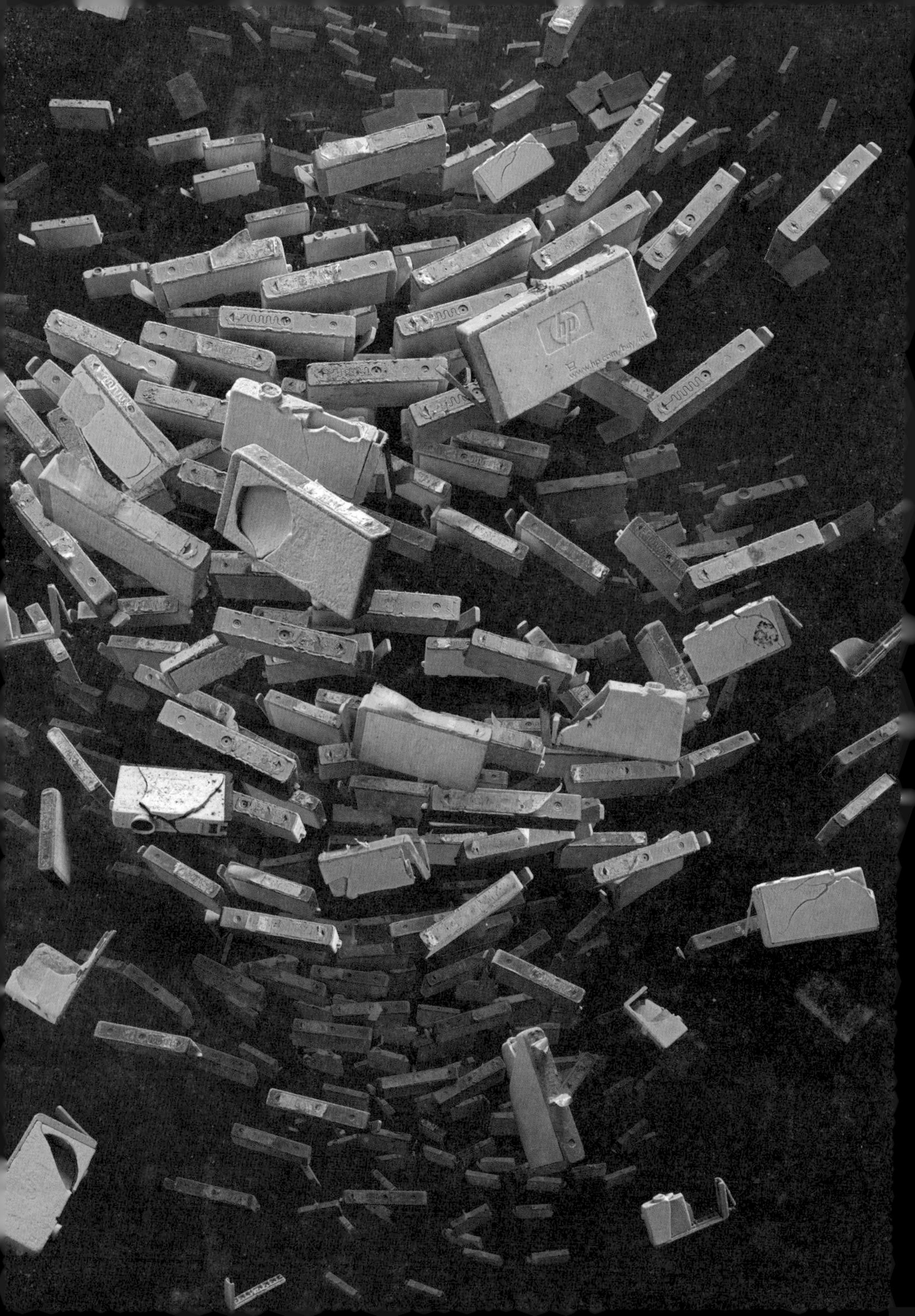
hp

No one knows how many goods from cargo spills are lying on the ocean floor. No one really knows how many items from shipping containers fall overboard every year. Currently, lost goods only have to be declared if they are classed as hazardous.

Companies' responses to cargo spills vary wildly. Some are full of remorse, keen to make amends. They fund beach cleans, offer to send teams of people to pick up lost cargo from beaches, set up hot lines and provide recycling facilities. Others keep silent, perhaps not wanting the embarrassment, inconvenience or legal implications of a cargo spill.

---

At any time, there are over 6,000 ships carrying containers around the world, huge vessels transporting 226 million giant boxes a year. Now stacked higher than ever before, they transport everything from trainers to televisions, from chairs to car tyres, from motorbikes to medical tape.

As Rose George says in her book, *Deep Sea and Foreign Going*, 90 per cent of everything we wear, we eat, we consume is brought to us by these huge floating warehouses. Some carry over 23,000 containers at a time.

However, not all make it safely to their destination. In its 2020 report, the World Shipping Council revealed an average of 1,382 containers had fallen overboard each year for the past twelve years. That figure was calculated before over 3,000 containers fell into the Pacific Ocean between October 2020 and January 2021, though.

*Printer cartridges from a shipping container lost overboard in the Atlantic Ocean in 2014, later washed up on coastlines from Florida to northern Norway. The incident enabled scientists to map how debris from a cargo spill is swept across the globe by ocean currents. Image © Mandy Barker.*

Not all sink straight away. Some carry on floating, posing a hazard to other vessels before finally drifting to the seabed. While many burst open on impact, others remain intact, the goods inside only escaping as the giant boxes are distorted by pounding waves and as saltwater eats away at the metal, turning it to rust.

Once, after a friend and I spent days picking up packets of soggy rice cakes washed ashore after four shipping containers fell from a cargo ship into the Irish Sea during a storm, a marine surveyor representing the vessel's owner emailed me: 'We look forward to working with you over the coming days/weeks to resolve this unfortunate issue and to see the beaches restored to their normal condition for all to enjoy.'

However, there's no mention of what happened to all the other goods lost overboard at the same time, the cargo that sank to the bottom of the sea. The flat pack furniture, the tool shop equipment, the wine, the packaging that went with them and the shipping containers themselves. What about them? Will the sea ever be restored to its normal condition?

You never really hear about the cargo that sinks to the sea floor. You only ever hear about the goods that wash ashore.

# SOME FAMOUS – *AND NOT SO FAMOUS* – CARGO SPILLS

In the eighteenth century, a cargo ship carrying earthenware beads was wrecked near St Agnes, on the Isles of Scilly. Many washed up at Wingletang Bay, now known locally as Beady Pool. The beads still occasionally turn up.

Bright orange Garfield phones have been littering the Iroise coast of Brittany since the 1980s. Beach cleaners had long suspected they might be coming from a lost shipping container but were unsure where it was. The mystery was eventually solved when a local farmer remembered the first Garfield phone washing up after a storm, which led campaigners to the secluded sea cave where the container was lodged.

In June 2013, the MOL *Comfort* broke in two in the Indian Ocean, shedding some 4,300 containers, the most ever lost in a single incident. The vessel was said to be transporting consumer electronics and clothing.

Thousands of plastic detergent bottles washed up in Cornwall in January 2016, turning the white crests of the waves bright pink as they floated ashore. It's thought they were from the MV *Blue Ocean*, which had lost a container holding nearly 19,000 near Land's End the previous year.

Hundreds of unworn trainers, flip-flops and sandals washed up on beaches from the Bahamas and Bermuda to Ireland and Scotland after the Maersk *Shanghai* lost over seventy containers off the North Carolina coast in March 2018. The runaway cargo also included kneepads and packets of figs.

Flat screen televisions, car parts, clothing, IKEA furniture and children's toys were among thousands of goods washed up on the Dutch islands of Terschelling and Vlieland and the German island of Borkum after the cargo ship MSC *Zoe* lost 342 containers in the North Sea in January 2019. At the time, it was reported many items, including little plastic ponies, could still be adrift. These were found on the shores of Terschelling by seven-year-old Jelle Budding. After collecting them up, he sold many at a local flea market, donating the money he raised to the Association for the Protection of the Wadden Sea.

*Festive flotsam: Christmas lighting decorations from a cargo spill have been discovered along the coasts of the UK and Ireland, the Channel Islands, France and the Canary Islands, spread far and wide by ocean currents. As well as Father Christmas designs, beachcombers have found shooting stars, Christmas trees, reindeer and sleighs, herald angels, candles and snowflakes.*

Thought to be from a cargo spill long ago, these lion, reindeer and gorilla plastic toys washed ashore in Cornwall, but beachcombers in Wales and France have found them too.

On 30 November 2020, the boxship *ONE Apus* lost over 1,800 containers when it hit severe weather in the Pacific Ocean. Fireworks, batteries and liquid ethanol were said to have been in some of the dangerous goods containers that fell overboard.

Toy wheels and the occasional toy roof rack have rolled up on beaches all the way from Cornwall to Wales. Fishermen have found parts from toy safari trucks in their nets too. Where they're coming from, we're not sure.

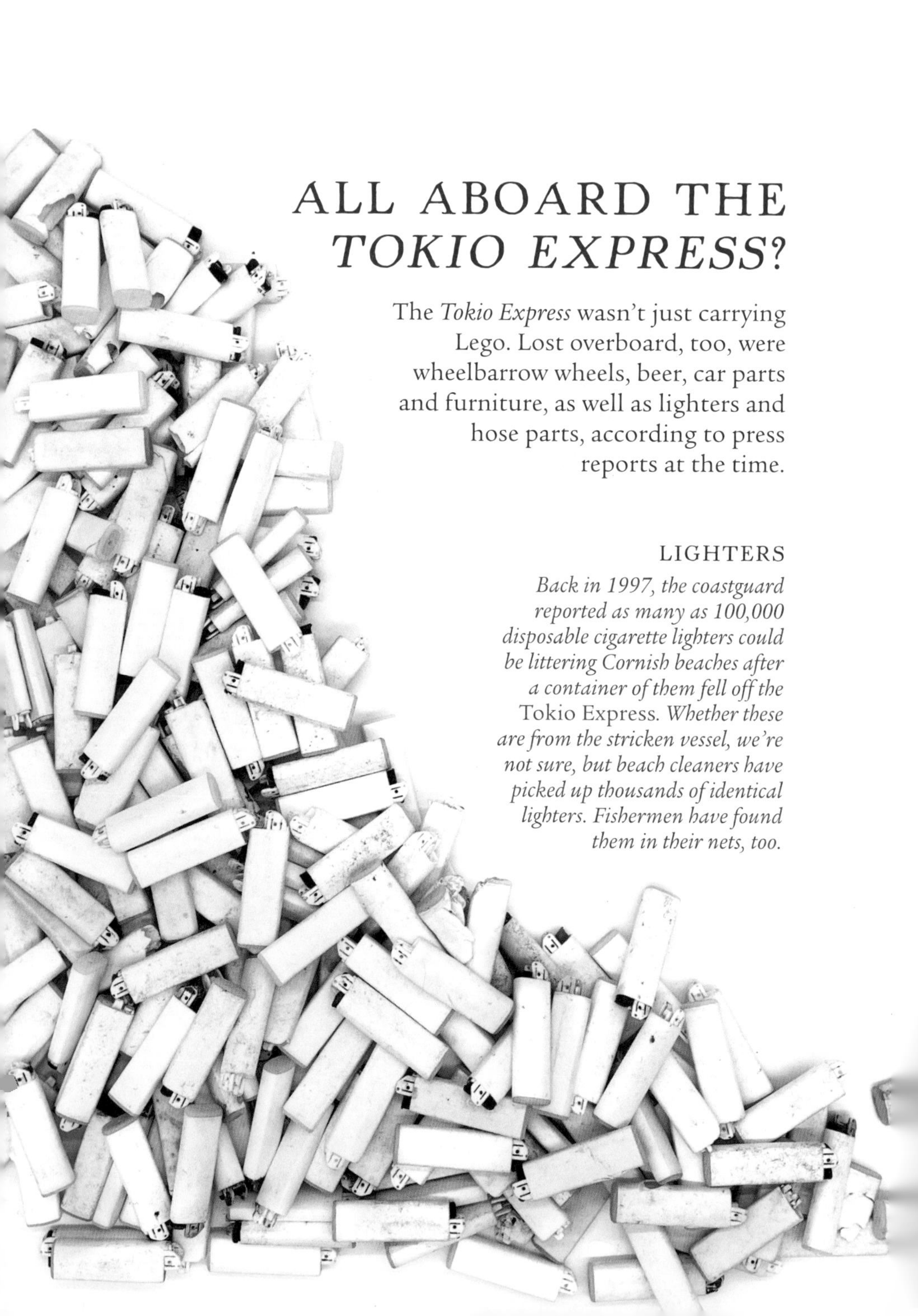

# ALL ABOARD THE *TOKIO EXPRESS*?

The *Tokio Express* wasn't just carrying Lego. Lost overboard, too, were wheelbarrow wheels, beer, car parts and furniture, as well as lighters and hose parts, according to press reports at the time.

## LIGHTERS

*Back in 1997, the coastguard reported as many as 100,000 disposable cigarette lighters could be littering Cornish beaches after a container of them fell off the* Tokio Express. *Whether these are from the stricken vessel, we're not sure, but beach cleaners have picked up thousands of identical lighters. Fishermen have found them in their nets, too.*

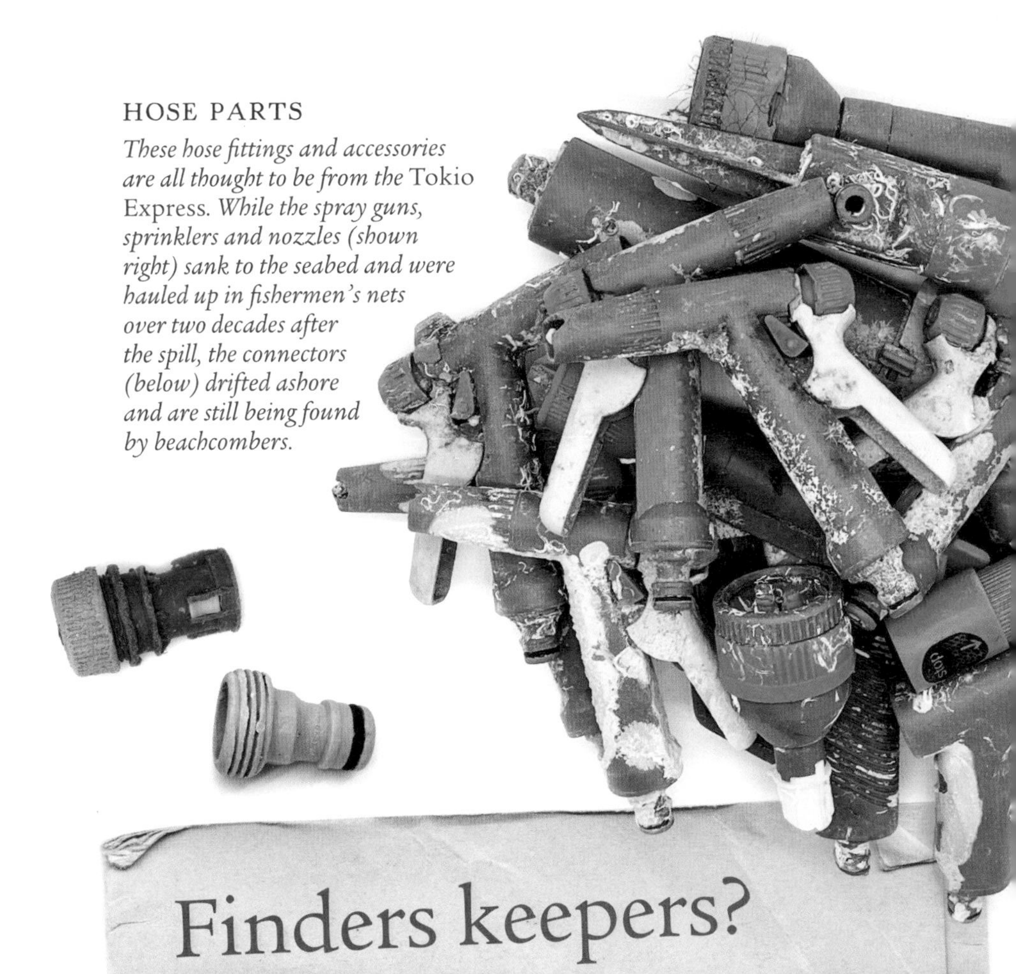

HOSE PARTS

*These hose fittings and accessories are all thought to be from the* Tokio Express. *While the spray guns, sprinklers and nozzles (shown right) sank to the seabed and were hauled up in fishermen's nets over two decades after the spill, the connectors (below) drifted ashore and are still being found by beachcombers.*

# Finders keepers?

If you find a piece of Lego on the beach, is it a case of finders keepers? Not necessarily. In the UK, all items thought to be from a cargo spill or shipwreck should be reported to the Maritime and Coastguard Agency's Receiver of Wreck.* It is their job to determine to whom the cargo belongs.

Did you know that the Receiver of Wreck is also responsible for managing 'royal fish' strandings in England and Wales? It's an ancient right dating back to Edward II's reign. In days gone by, whales, porpoises, dolphins and sturgeon were considered a lucrative haul by wreckers, as oil and other products could be salvaged from the carcass.

*See page 176 for details of the Receiver of Wreck*

# Sea Glass

Bernadette Noll

I want to age like sea glass.
Smoothed by tides, not broken.
I want the currents of life to toss me around,
shake me up and leave me feeling washed clean.
I want my hard edges to soften as the years pass –
made not weak, but supple.
I want to ride the waves,
go with the flow,
feel the impact of the surging tides rolling in and out.

# Archaeology of the Tides

## *Artefacts of the Plastic Age*

# 7

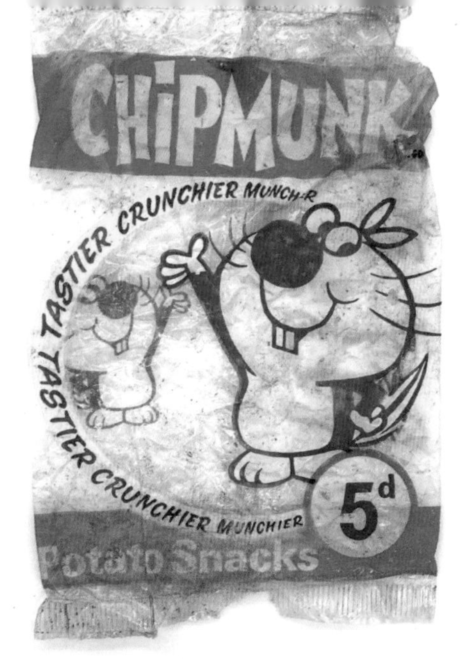


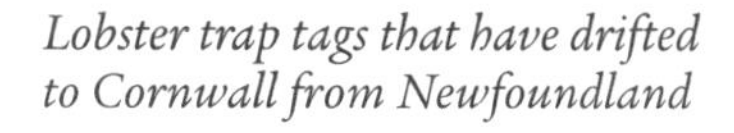


*Lobster trap tags that have drifted to Cornwall from Newfoundland*

*Old crisp packets found during beach cleans – Chipmunk Potato Snacks from 1971 or before and Smiths Crisps from 1980. Not adrift in the ocean for decades but exposed recently by shifting sands.*

*Plastic drift card found on a beach in North Cornwall in 2014 by beachcomber Chris Easton. Research by Dr Ebbesmeyer revealed it had been released by American scientific agency, National Oceanic and Atmospheric Administration (NOAA), in 1976 on Nantucket Island, Massachusetts, USA to track oil spilled from the tanker* Argo Merchant. *The vessel had run aground and broken apart, resulting in one of the largest marine oil spills in history. The time elapsed between the card's release and its discovery 3,000 miles away was roughly thirty-seven years and two months, believed to be a record for these drift cards. The card probably wasn't at sea all that time though. It's thought it may have washed ashore decades earlier and become buried in wind-blown sand.*

It was a faded and abraded plastic disc with the faint words 'Kellogg's – The Greatest Name in Cereals' that first made me wonder how old some of the plastic found on beaches might be. I'd discovered it among swathes of rubbish strewn across the sand after a winter storm. It looked old, but how old?

Once back home I searched online and came across the website cerealoffers.com, run by the extremely knowledgeable and helpful Nick Symes. Among thousands of pictures of free toys given away in cereal packets, as well as adverts and promotional items relating to Weetabix, Quaker and Kellogg's, I found my disc. It was the remains of a football spinning top, free inside packets of Kellogg's Corn Flakes in 1958.

Not all debris found on beaches is freshly washed in from the sea. During storms and high spring tides, pounding waves eat into dunes and sandbanks, releasing plastic that has been trapped for decades.

Some will have been left by beachgoers long ago: toys, crisp packets, cola cans and carrier bags – remnants of picnics past. Much, though, will have been

*Plastic exposed after storm-driven waves have eaten into the dunes, leaving a vertical wall of sand*

brought in by the sea years earlier, often becoming buried in wind-blown sand during gales: thirty-year-old lobster trap tags that have drifted thousands of miles from Canada and the US, spirit bottles from Denmark, washing-up bottles from Poland and bottles of hair dye from Japan, perhaps from passing ships.

Dr Þóra Pétursdóttir, associate professor of archaeology at the University of Oslo, who has been studying flotsam washed up on remote beaches in northern Norway and Iceland for years, describes this seaborne debris as drift matter, saying it must be taken seriously in archaeology.

'Although these objects can't always be traced to a specific place or culture, many have already become part of the environment,' she says. 'They might not have any connection to the beach where they were found but that doesn't mean they shouldn't be of interest to archaeologists.

'While of course both harmful and problematic, this seaborne debris is still our heritage. It just calls for a different approach. Drift matter evokes a museum collection of the future, not the past.'

After discovering the football spinning top was over fifty years old, I became intrigued by the age of some of the other plastic items picked up from the beach – classic Lego bricks, ancient cereal packet figurines, faded flowers, little green army men, Wild West figures and the tiny farm animals.

Cataloguing them began as a bit of a joke with friends working in archaeology. They would send pictures of their finds, such as flint tools, Iron Age coins, Saxon grave markers and seventeenth-century clay pipes. In response, I would send them a picture of 'Barbie's boudoir mirror, circa 1964', or a 'Zorro ring, free in packets of Puffed Wheat in 1959' (left).

But as time went on, I realised much of the plastic I'd found was far older than at first thought. Not just the toys, but plugs made from Bakelite, poppit beads from the 1950s, 'buried treasure' ice-cream sticks from the 1960s and 1970s, Noddy toothbrushes and old-fashioned hair curlers, which hit me with a wave of nostalgia every time I came across one.

Some of the curtain rings I'd picked up were marked Roanoid, the trade name for a plastic first developed in 1923 and used for hooks and handles on the ocean liner *Queen Mary* in 1936.

Even the yellowing fragments of old vinyl safety goggles I'd fished out from the bottom of a rock pool seemed to be vintage, bearing names such as Vizorette and Panoramette, brands that had their heyday in the 1950s.

And I began to wonder whether these plastics of the past would perhaps in future no longer be seen as just beach litter but as artefacts to be studied closely by archaeologists and historians.

*Super Doby washing-up bottle thought to date from 1955–9*

*Baking powder submarines* – first given away in packets of Kellogg's Corn Flakes in 1957. In later years, they also came free with Frosties, Coco Pops, Sugar Smacks, Ricicles and All Stars. While the yellow subs here are thought to date to 1986, the orange one could either be from 1957 or 1963, when they were re-issued.

*Robin Hood* – free inside packets of Kellogg's Corn Flakes in the 1960s.

1950s

1960s

*Wild west figure** – given away in packets of Sugar Puffs in 1957.

*Native American figure** – given away free in packets of Sugar Puffs in 1957.

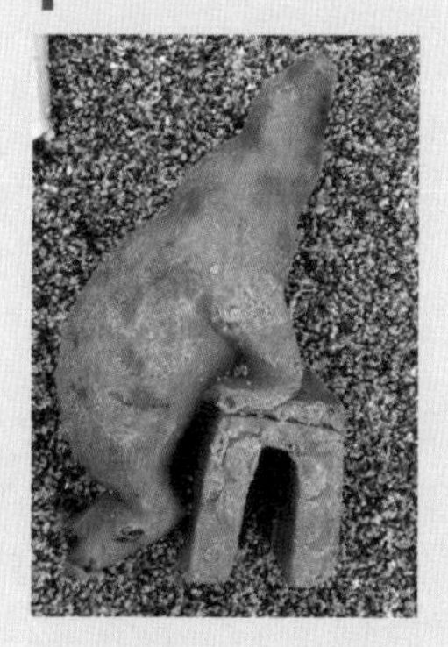

*Sea lion** – given away with Kellogg's Sugar Smacks in 1959. From the Circus Models range.

**Also available commercially.*

# *Archaeology of the Cereal Packet Toy*

*Big Ears* – given away with Kellogg's Ricicles in 1962 and 1967.

*Cockerel cycle reflector* – free inside boxes of Kellogg's Corn Flakes in 1989.

1970s 1980s 1990s

*Giraffe** – free with packets of Kellogg's Puffa Puffa Rice and Coco Krispies in 1971.

*Clip-together Gordon Banks spot kicker* – free with Shredded Wheat in 1974.

*Sporting Tony the Tiger* – given away with Kellogg's Frosties in 1991.

# ARTEFACTS OF THE PLASTIC AGE

Some washed out of the sand after storms, some washed in from the sea floor.

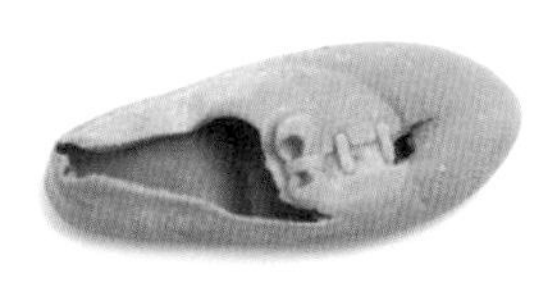

## DOLL'S SHOE

a lime green flat lace-up shoe belonging to Sindy's French friend Mitzi, introduced in autumn 1967

## CLUTCH BAG WITH A PEARL CLASP

From Sindy's Dream Date, 1963 and found washed ashore with other plastic from the seabed in 2020

## BARBIE'S BOUDOIR MIRROR c. 1964

## HAX BOTTLE

Strawberry Flip Syrup bottle bearing the name HAX. According to the Museum of Design in Plastics, these fruit-shaped HAX bottles are loosely dated from 1950–69

## GI JOE FATIGUE CAP

made in Canada by Hasbro
between 1965 and 1970

## PLASTIC JET CAR

by Tallon, a Bristol-based company that also sold stationery, pens, jigsaws and boxed games. This car appeared in its 1962 catalogue, 'Hours of fun for only a shilling'

## MERCEDEZ-BENZ CAR

Made by Tomte Laerdal of Stavanger, Norway between 1962 and 1973

## ICE SKATER FIGURINE*

thought to date to the 1950s

## FIRE EXTINGUISHER

from Evel Knievel's Scramble Van, 1973

## NATIVE AMERICAN WOMAN WITH INFANT

Made by toy company Britains, this version was probably produced between 1965 and 1976

** Found by Neil Redfern of the Council for British Archaeology on a beach clean to celebrate the Festival of Archaeology in July 2021.*

# BAKELITE PLUG

The first plastic produced from synthetic components, Bakelite was invented in the early twentieth century but this plug is believed to date from **1960–66**

# FISH TAG WITH 'LETTER INSIDE'

## C. 1970

Not much more than an inch long, this tiny 'message in a bottle' is from a fish tagging experiment thought to have taken place in about 1970. It was originally attached to a fish by scientists at the Polar Research Institute of Marine Fisheries and Oceanography in Murmansk, Russia.

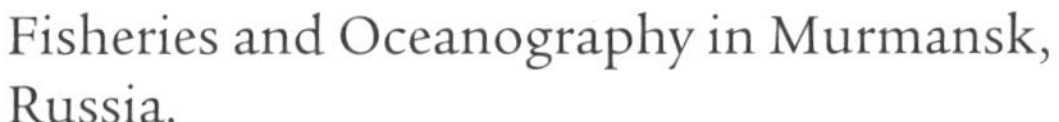

Inside the plastic capsule was a faded slip of paper with a message written in Russian and English. Offering a reward, it instructs the finder to send details of the fish, including its length and where and when it was captured, to the UK's Lowestoft Fisheries Laboratory, now known as CEFAS. The tiny letter also asks the finder to send the fish's otoliths (or ear stones), which help scientists determine its age. No one knows what happened to the fish, but some forty-five years after it was tagged the capsule was discovered on a Cornish beach.

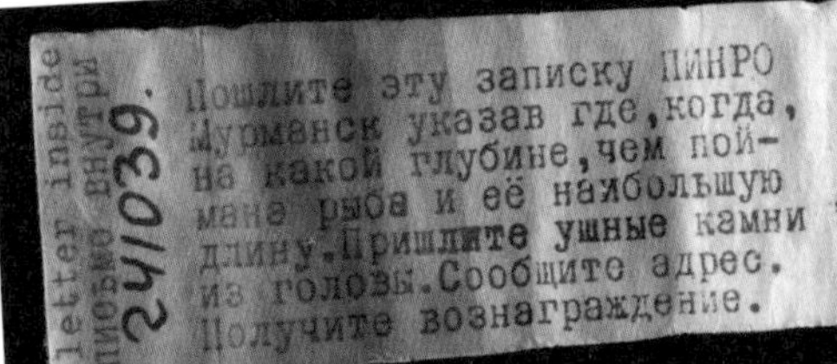

letter inside
письмо ВНУТРИ
24/039.
Пошлите эту записку ПИНРО
Мурманск указав где, когда,
на какой глубине, чем пой-
мана рыба и её наибольшую
длину. Пришлите ушные камни
из головы. Сообщите адрес.
Получите вознаграждение.

USSR
24/039
Please se this slip to
Fisheries Laboratory, Lowestof
England with your name and
address, date place and depth
capture. If possible, state tot
length of fish and send ear-
stones. Reward.

# THE LINGERING LEGACY OF A LOLLY STICK

*moments to lick,*
*decades buried in the sand...*

Part of a Wall's Count Dracula's Secret 'Trace-A-Face' lolly stick from the 1970s. These had shapes such as eyes, noses and mouths cut into them to make a stencil.

Pokémon lollipop sticks, thought to date to 1999.

Buried Treasure Ice Cream Sticks, produced in the USA from the early 1960s through to the mid-1970s. There were sixty-five different characters, but you never knew which one you would get until you had finished the ice cream.

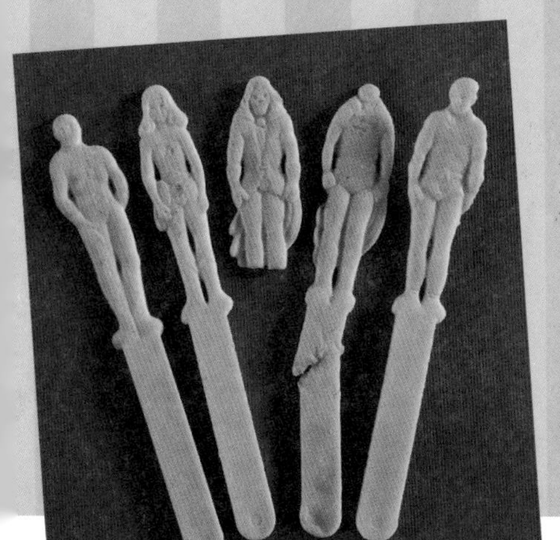

Lyons Maid Superman's Superheroes plastic lolly sticks from the 1970s. There were twelve sticks to collect: six heroes (Superman, Wonder Woman, Batman, Flash, Robin and Aquaman) and six villains (Catwoman, The Riddler, Captain Cold, The Penguin, The Joker and Doctor Destiny).

# SURF SAFARI

Where do they all come from, these tiny plastic animals saved from the waves?

While some will have been lost by children playing at the beach, others, such as the lions, reindeer and gorilla, are thought to be from a cargo spill and have washed up in the UK and in France.

The little black bull would once have been attached by ribbon to a bottle of Spanish Torres Sangre de Toro wine, a gimmick the winery discontinued in 2012 to reduce its use of plastic and minimise its impact on the environment.

Some of the animals are from popular games, such as the mouse from Mouse Trap, first produced in 1963. The monkey is from the Barrel of Monkeys game, which made its debut in 1965, while the bug is from Bedbugs, originally released in 1985.

The miniature white horse token would have hung around the neck of a bottle of White Horse Whisky and was first introduced in the 1960s, while the cougar is from a Kinder Surprise egg found unopened on the beach.

Others could be from lucky dip and jamboree bags, which often had toys and charms inside, or from Lucky Tatties, white fondant sweets coated in cinnamon that came with a trinket. A few may be from gumball machines, crackers and comics.

Plastic animals were also given away in cereal packets. The performing sea lion circus toy was free inside Kellogg's Sugar Smacks in 1959, while the giraffe could be found in Kellogg's Coco Krispies in 1971.

# OCEAN ARMY

Plastic toy soldiers are a surprisingly common beach find.

Little plastic army men were first produced in the USA in the 1930s, making their debut in the UK in the 1940s. In 1958, model guards bandsmen made in 'unbreakable scarlet plastic' were given away free in packets of Kellogg's Corn Flakes. During the 1960s, OO scale Airfix soldiers could be found in Sugar Puffs, three in a family-sized packet, two in a standard size.

These days you can buy toy soldiers by the bucketful. They're even used in aquariums as biofilters – surfaces on which micro-organisms can grow. That might explain why so many end up in the sea, perhaps tipped out with dirty water into drains.

Some of the toy soldiers here are thought to date back to the 1950s. There are no women warriors in this collection, though. Little plastic army women weren't introduced until 2020.

ROWN TREE
smarties smarties

# ANCIENT 'SMARTEFACTS'

Plastic Smartie lids regularly turn up on beaches, despite the fact they were phased out in 2005.

When the Smarties brand was first launched in 1937, the colourful chocolates came in cylinder-shaped packets but in or around 1960, the Smartie tube with a removable plastic lid was introduced.

Early Smartie lids had 'ROWNTREE' on top, with a capital letter or number on the reverse. In 1971, the capital letters and numbers were replaced with lower case letters to reflect the style of lettering children were taught in schools. These days, lids with capital letters and numbers are rare and much sought-after by collectors.

In 1988, Swiss-based Nestlé bought the company and the Rowntree name on the lid was replaced by the word 'Smarties'. In 2005, Nestlé Rowntree announced it was ditching the plastic lid altogether and replacing the old cylindrical design with a hexagonal or six-sided pack with a cardboard flip-top.

In March 2020, we picked up over seventy plastic Smartie lids from the same beach over twenty-four hours. Some were nearly sixty years old.

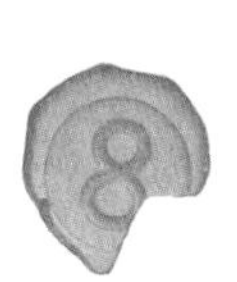

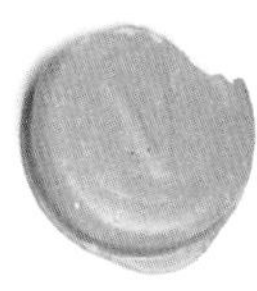

# A (PLASTIC) PEBBLE IN YOUR POCKET?

Smooth, sea-worn pebbles. They might look like the real thing but these are plastic pebbles, created when plastic is burnt. At first glance you might not even notice them as you walk along the beach. Often, it's their position on the sand that gives them away. Whereas natural rocks and pebbles tend to be scattered across a beach, these 'pyroplastic' pebbles float and are usually found on the high tide mark with the rest of the plastic washed ashore. If you pick them up, they feel different too, lighter and warmer to the touch than natural rocks. It's not entirely clear where they are all coming from; perhaps from old landfill sites eroding into the sea, from plastic burnt and thrown into the ocean, from campfires and beach bonfires.

When scientists from the University of Plymouth analysed a selection of plastic 'stones' picked up from Cornish beaches to see what they were made from, they discovered all contained polyethylene and polypropylene, plastics commonly used in carrier bags, food packaging and plant pots. This plastic pebble (left) has the remains of a piece of Lego sea grass from the cargo spill embedded in it. When melted plastic combines with sand, rocks and wood, it's known as plastiglomerate.

There are two real pebbles among the plastic ones in the picture on the right. Can you tell which they are? *(Answer on p. 177)*

# AMBERGRIS, TREASURE OF THE SEA

Sometimes there's an unexpected bonus to picking up plastic pebbles. On a windswept beach early in 2020 I met up with a beachcomber who hunts for ambergris, the waxy substance that originates in the intestines of sperm whales. Highly prized in perfume making, ambergris is known as 'treasure of the sea', often fetching higher prices than gold. So I knew what to look out for, the beachcomber had generously given me several fragments of ambergris he kept in his pocket. Later, while rummaging through a box of plastic pebbles to find suitable samples to send to a school, I noticed one looked remarkably similar to the ambergris the beachcomber had given me. It was ashy and marbled, like pale stone, flecked with black. It was the distinctive smell, though, that made me think it might be ambergris. When fresh, ambergris is said to have a strong, faecal odour. As it dries out, the smell is said to improve, becoming sweet, earthy and musky. This was somewhere in between. Later, I met up with an ambergris expert who confirmed my 'plastic pebble' was indeed floating gold, as it is known. Not enough to make my fortune, but an unexpected addition to my cabinet of beachcombed curiosities.

*In ancient China, ambergris was thought to be dragon's spittle hardened by the sea*

## MONOPOLY HOUSES AND HOTELS

It's a mystery why so many plastic Monopoly houses and hotels wash ashore. Many beachcombers have found them. It's tempting to think they're from a cargo spill, as they turn up on both sides of the Atlantic as well as Pacific coasts but seem to date to different periods. During research for her book *Rag & Bone*, Lisa Woollett discovered plain plastic houses were introduced in the 1960s, with overhanging roofs added in the 1970s and chimneys in the 1980s.

## CAKES, TRINKETS AND PLASTIC BABIES

This little white plastic figurine or fève would have been hidden inside a French cake known as a galette des rois or king cake, traditionally baked for Epiphany on 6 January. Whoever found the fève in their slice of cake would become king or queen for the day. In an American king cake, usually eaten during Mardi Gras in New Orleans, the fève is often a small plastic baby, symbolising baby Jesus, though whether these are king cake babies, charms or lost toys I am not sure.

# SHELL SHOCKED

Real shells, plastic shells, all found on Cornish beaches. Can you tell which is which? Some of the shells here are the plastic cases of a lickable sweet known as Schleckmuscheln, first introduced in the 1960s. *(Answer on page 177)*

# FAKE FLOWERS AND FOLIAGE

Plastic flowers and fake foliage are another common beach find.

Walk along a clifftop and you'll often see bunches of artificial flowers laid on the grass, placed on memorial benches and tied to railings. Many end up in the sea as gale force winds scatter them far and wide.

Some of the flowers are thought to date back to the 1960s and 1970s, when millions of plastic daffodils and tulips made in the Far East were given away with packets of Persil by shopkeepers in the soap powder wars between Procter & Gamble and Lever Brothers.

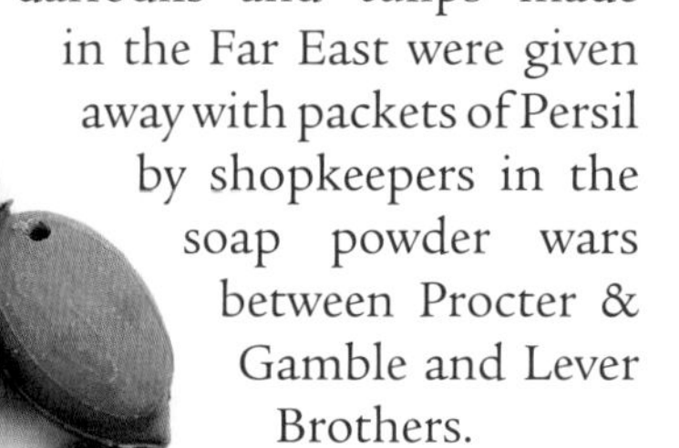

Washed up, too, are aquarium plants, perhaps thrown out with dirty water from fish tanks, and red and green leaves used in butchers' and fishmongers' shops as 'garnish and dividers'.

Among the fake foliage are sprigs of 'box' from artificial trees and hanging topiary balls found outside houses and shops in towns and cities, plastic pine needles from Christmas trees and festive wreaths and plastic stems from remembrance poppies. Many find their way into the ocean after falling on to pavements and being washed into storm drains after heavy rain.

As for all the plastic walnuts, almonds and hazelnuts that have been washing up for years, where do they come from? Are they from a cargo spill long ago?

# BROKEN BUCKETS, IN SPADES

Lost, forgotten and discarded sand toys are a big problem on beaches, particularly in the summer months. Sold by the bagful at seafront shops, supermarkets, petrol stations and pound stores, they're regularly abandoned by beachgoers, along with cheap polystyrene bodyboards, spades and flimsy crabbing buckets that invariably break on their first outing. The mesh bags they come in, and the plastic clips securing them, are often discarded at the beach too, adding to the litter problem and posing a hazard to wildlife.

In 2021, we picked up over 400 brightly coloured beach toys and fishing nets from two beaches on the north coast of Cornwall over the course of the summer. Many had been shipped halfway round the world, only to break the first time they were used.

## WHY ARE THERE SO MANY TOOTHBRUSHES IN THE SEA?

All these plastic toothbrushes were picked up during beach cleans. While a few are new and thought to be from a cargo spill, most have been used.

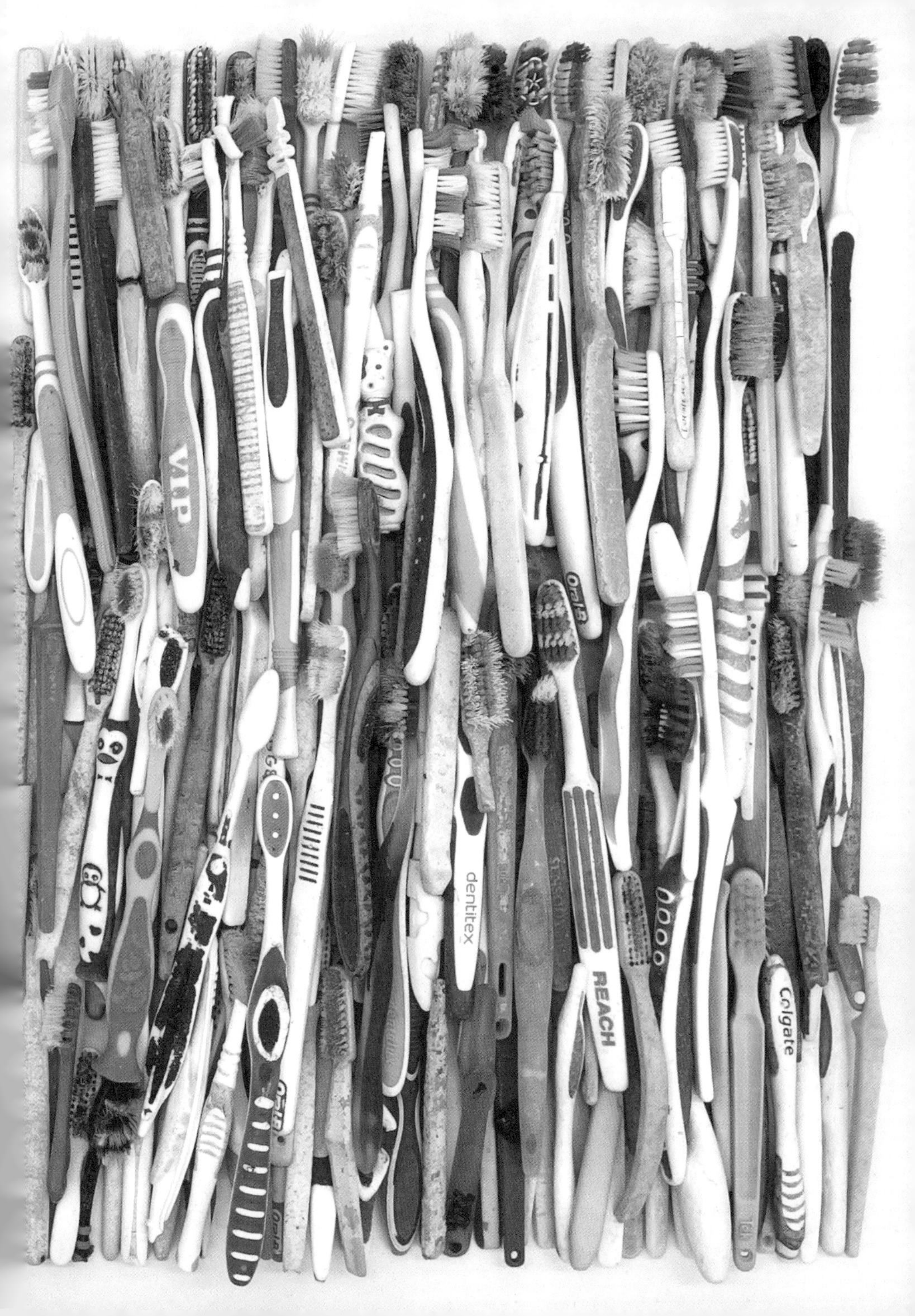
VIP
Oral B
dentitex
REACH
Colgate

# 8

# When the tide goes out

## *A beachcomber's dream or an environmentalist's nightmare?*

And till my ghastly tale is told,
This heart within me burns.

extract from
The Rime of the Ancient Mariner
SAMUEL TAYLOR COLERIDGE

# 'LEGO BRICKS BELONG AT HOME, NOT IN THE OCEAN'

## *An Interview with Lego*

'We don't want any Lego bricks to be in the sea – ever,' says Tim Brooks, Lego's vice president for environmental responsibility. 'We're absolutely passionate about keeping Lego bricks out of nature.'

Tim and I meet over a video call in August 2021 to talk about the lingering impact of the Lego spill.

We chat while he's in his office at Billund in Denmark and I'm at home in Cornwall, surrounded by ever-growing mounds of beach finds accumulated over the years.

As Tim points out, moving goods around the world can create difficult situations, especially when things go wrong and they end up slipping beneath the waves.

'We work with reputable shipping companies, but once a container is loaded on to the ship and it sails, we can't control what happens next,' he says. 'We rely on them to make sure they don't harm the environment.'

Tim talks about Lego's focus on environmental sustainability and the responsible use of plastic, to encourage children and families to keep their bricks in play so they don't enter landfill or get lost in the environment.

'Although the 1997 Lego spill is a historic case, it does highlight the importance of this issue,' he says.

While the initial clean-up of cargo from the *Tokio Express* was paid for by the vessel's insurers, Lego provided a detailed list of bricks lost, along with samples and polaroid pictures of the boxes used to transport them, so beachcombers knew what to look out for. Children who wrote in to say they'd found bricks from the spill received free underwater explorer sets.

'One of the things we're most proud of is that Lego is compatible and long lasting,' Tim says. 'There's no logical reason why you would ever throw a Lego brick out. They're safe, durable and can be handed down from generation to generation. Plastic is a great material but it's great in context. It definitely belongs in the home or in schools, not in the ocean.'

These days it's rare for Lego bricks to be sent by sea, though, as Tim points out, it is still the least polluting way to ship goods to the corners of the globe.

'Nowadays we tend to manufacture Lego in our local markets. A third of our bricks are made here in Billund in Denmark and transported by road to Europe. We have a

factory in Mexico to serve the mainland US and another in China for the Asian market. We don't often transport by sea unless we have a particular shortage in a certain area.'

Lego is working hard to build its bricks from more sustainable raw materials. In 2018 it introduced its first sustainable elements, including green leaves, bushes and trees made from sugarcane-based polyethylene instead of oil-based plastic.

It's phasing out single-use disposable plastic from its packaging to make it sustainable by 2025.

To keep its bricks in circulation for as long as possible, Lego has launched Replay, a take back scheme that inspires owners to pass on bricks they no longer use to children who need them most. So far, the scheme has been trialled in the United States and Canada but there are plans to roll it out to more countries in the coming years.

In 2021, Lego unveiled a prototype brick made from recycled plastic, the latest step in its journey to make its products from sustainable materials.

As Tim says, Lego 'can't turn back the clock, much as we'd like to' to prevent the cargo spill from the *Tokio Express*. And while the company is working on ways of making its bricks more eco-friendly in future, it cannot change what happened on that fateful day.

'I'd love to go back twenty-five years and not let anything get in the ocean,' he says.

JESU
123

Inspired by Dr Ebbesmeyer's earlier experiments, I met up with environmental scientist Dr Andrew Turner to see if we could determine once and for all which pieces of Lego from the spill floated, which sank and why beachcombers had never found many of the bricks thought to have been cast adrift.

Our original plans to carry out experiments in a laboratory to determine what type of material the bricks were made from, and therefore whether they would be likely to float or sink in the ocean, were scuppered by the Covid-19 pandemic. So armed with a box of Lego and a bowl of seawater, we opted for a spot of backyard oceanography instead.

When Dr Ebbesmeyer carried out his original experiments in 1997, he predicted a significant number of the lost Lego elements would probably float, so long as they were adrift. Because he was concerned the paper labels stuck to the samples might come off, he didn't leave the bricks in water for long.

In Dr Turner's garden, we repeated the exercise, this time using identical pieces of Lego without paper labels to see what happened when they were immersed in water for longer.

Not surprisingly, we discovered the Lego picked up regularly by beachcombers, including the octopuses, scuba tanks, brooms, sea grass, cutlasses, life jackets, spear guns, flippers, ship's rigging and flowers, all floated. Because these are made of lightweight plastics, they are buoyant in seawater.

However, in our experiments nearly all the other Lego pieces sank, although not immediately. While some plummeted to the bottom like stones, a few drifted downwards before briefly bobbing back up again. And some floated – until the air inside them was expelled.

Lego elements are made from many different types of plastic. Classic bricks are produced from a substance known as ABS (acrylonitrile butadiene styrene), a hard plastic denser than water. Another dense plastic was also once used for transparent Lego pieces, such as windscreens.

When air bubbles cling to the surface or become trapped inside, they float. But, as we discovered, if you tap the bricks to dislodge the bubbles or poke the air out, they sink. Water temperature and the amount of salt in the water affect buoyancy too.

Our experiments weren't entirely conclusive and only intended as a rough continuation of the earlier research. But they could go some way to explaining why millions of pieces of Lego from the spill have never been found. The big unknown is whether some could still be trapped in plastic boxes or in the remains of the shipping container, though that may long since have rusted away.

As for the 28,700 little yellow life rafts, did any cross the Atlantic? No one knows for sure. Although one drifted ashore in Maine on the east coast of the USA, it's impossible to tell if it was from the spill.

When new, the life rafts have pockets of air inside that keep them afloat. So do the Lego dragons. But when the air escapes or is forced out, they sink. And while some life rafts have made landfall over the years, many have turned up in fishermen's nets, hauled up from the seabed over 20 miles off the Cornish coast.

Maybe there are still tens of thousands at the bottom of the sea. Some resting in sand, others drifting along the ocean floor, swept back and forth by underwater currents. A few making their way ashore with the rest of the debris caught up in the kelp, more heading into deeper waters. Many lost forever in the cold, silent deep.

In many ways, the curious tale of the Lego lost at sea is a story without an end, a vast jigsaw with millions of pieces still missing. An ocean mystery that fascinates some but horrifies others.

Where is all the Lego seldom seen – the sharks, the magic wands, the black bats and the witches' hats?

Once, in the fading light of a cold February afternoon, I noticed a strange brown disc wedged in a rock pool. Overnight, wild seas had stripped the rocks bare, exposing artefacts normally buried beneath tons of sand, including old ships' timbers still with their wooden pegs or trenails and ribs of twisted metal from a US bomber that crashed into rocks here in 1943.

At first, I assumed the disc was made of old plastic, perhaps Bakelite, and added it to the bag of rubbish I had picked up. Later that night, however, as I sorted through the plastic in the artificial light of my kitchen, the disk looked different, older.

Around the edge was a distinctive ring-and-dot pattern, one I recognised from time spent on archaeological digs. It was an ancient spindle whorl, once used for spinning fibres into yarn or thread. Not plastic or Bakelite, but bone.

*Spindle whorl – a disc or spherical object used in the ancient art of cloth making*

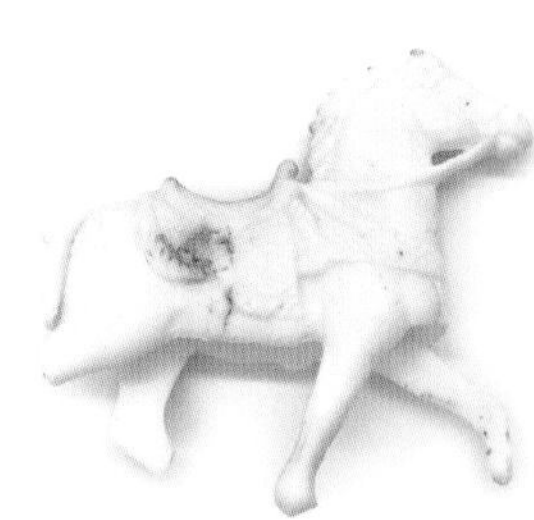

The local finds liaison officer from the Portable Antiquities Scheme subsequently dated it to the early medieval period, AD850–1100. Analysis of the bone showed it was probably from a red or fallow deer, perhaps even an elk, and may have come from Norway.

*Carved out of walrus ivory and sperm whale tooth in the late 12th to early 13th century, the Lewis Chessmen are the most famous chess pieces in the world. It's thought they were found in or before 1831 at Uig on the Isle of Lewis after a sandbank collapsed.*

It had lain undiscovered for 1,000 years, finally revealed by shifting sands. And it made me wonder if, hundreds of years from now, beachgoers will still be finding Lego.

Will a beachcomber of the future chance upon a hoard of dragons washed out of a sandbank, just as a crofter is said to have discovered the medieval Lewis chessmen in the nineteenth century in the Outer Hebrides of Scotland?

Maybe some of the Lego lying on the seabed will continue making its way ashore for years to come, swept hundreds of miles by sea floor currents.

Perhaps much will break down into microplastics and be ingested by fish and other marine creatures, including seabed-dwelling sharks.

Or maybe it will become buried in sand and sediment, eventually forming part of the geological record.

Sometimes I wonder if millions of years from now, a fossil hunter will take a hammer to a rock and discover a perfectly preserved Lego shark inside, or the trace of a Lego octopus.

A fossil of the Plastic Age.

A legacy of Lego.

For many people who pick up plastic from beaches, finding a bit of 'treasure' is what makes it fun – a Lego dragon, a toy soldier, a Smartie lid, a tiny dinosaur – but obviously it would be better if it wasn't there at all.

As eleven-year-old Laura, finder of a Lego octopus, said after one of our beach cleans many moons ago: 'I like to find treasure. But what I most like to find is a clean beach.'

COMMON ITEMS OF PLASTIC LITTER FOUND ON BEACHES INCLUDE:

- bags and wrappers
- bottles and drinks containers, including coffee cups
- food containers and cutlery
- smoking-related items (tobacco pouches, cigarette packets, stubs and lighters)
- synthetic rope and fishing-related items
- caps and lids
- industrial packaging
- polystyrene fragments
- sewage-related debris, including cotton bud sticks
- straws and stirrers
- sanitary items/wet wipes
- medical related items, including face masks and gloves

# *Beachcombers' bingo*

## *Here are some of the other items you might find – what do you recognise?*

*Answers on page 177*

# *Beachcombers' bingo 2*

## CARGO SPILL EDITION

*1*

*2*

*3*

*4*

*5*

*6*

*7*

*8*

9

10

*All of these items are thought to be from cargo spills...*

*See page 177 for more details*

11

12

13

14

15

16

17

# INGREDIENTS IN THE PLASTIC SOUP*

The items shown in the bowl were all scooped out of tidal pools on the north coast of Cornwall after storms. Around the outside are objects often found among the plastic on the strandline.

*Fishing beads – used in sea fishing and shore angling; many are luminous under UV light*

*Nurdles – the raw material from which most plastic goods are made, 2-5mm in diameter*

*Bio-beads – used by wastewater treatment plants to filter sewage*

*Airsoft pellets – spherical plastic projectiles used in airsoft guns*

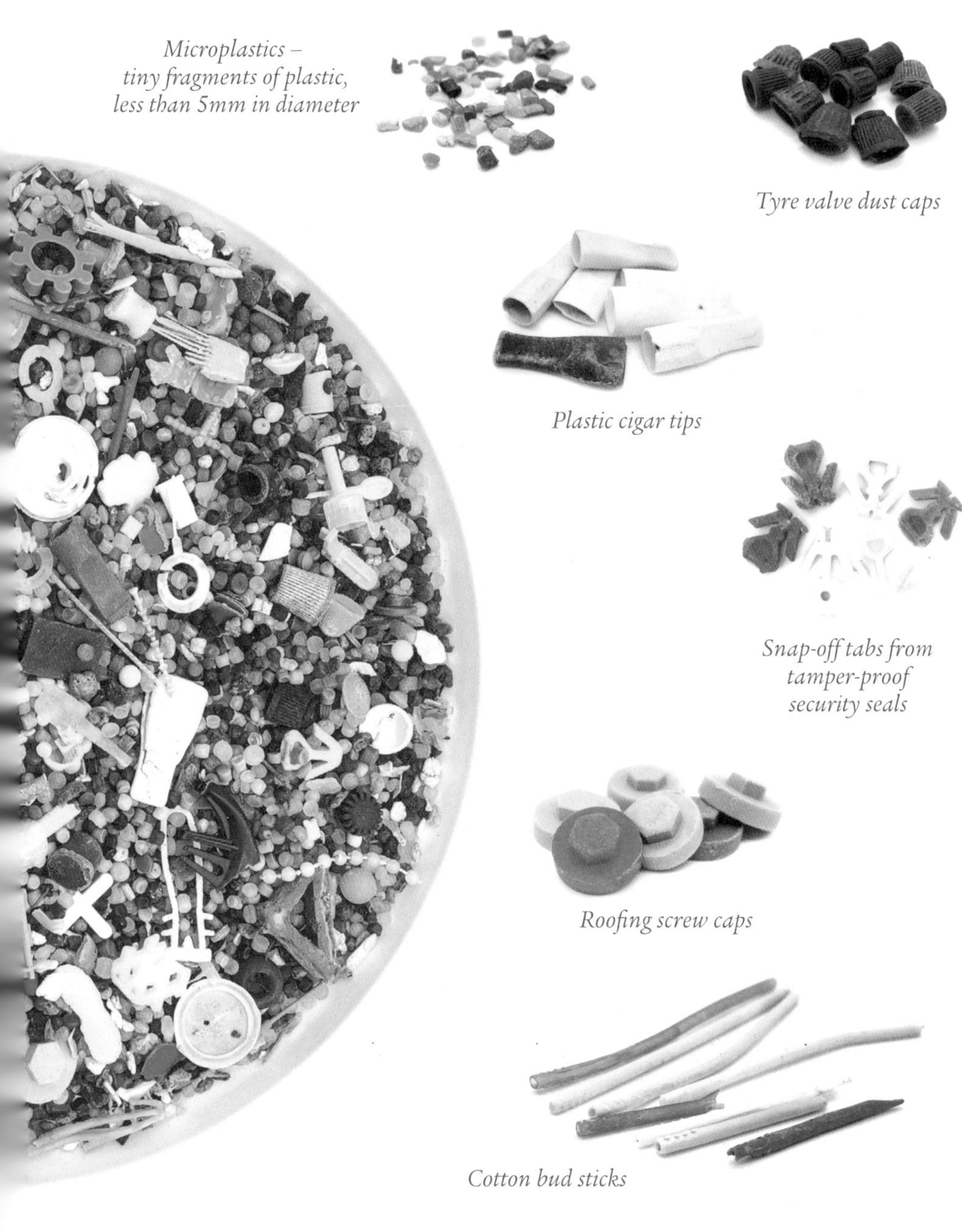

*Microplastics – tiny fragments of plastic, less than 5mm in diameter*

*Tyre valve dust caps*

*Plastic cigar tips*

*Snap-off tabs from tamper-proof security seals*

*Roofing screw caps*

*Cotton bud sticks*

** The term 'plastic soup' was coined by oceanographer and boat captain Charles Moore, who discovered the Great Pacific Garbage Patch. It refers to plastic waste found floating in the ocean.*

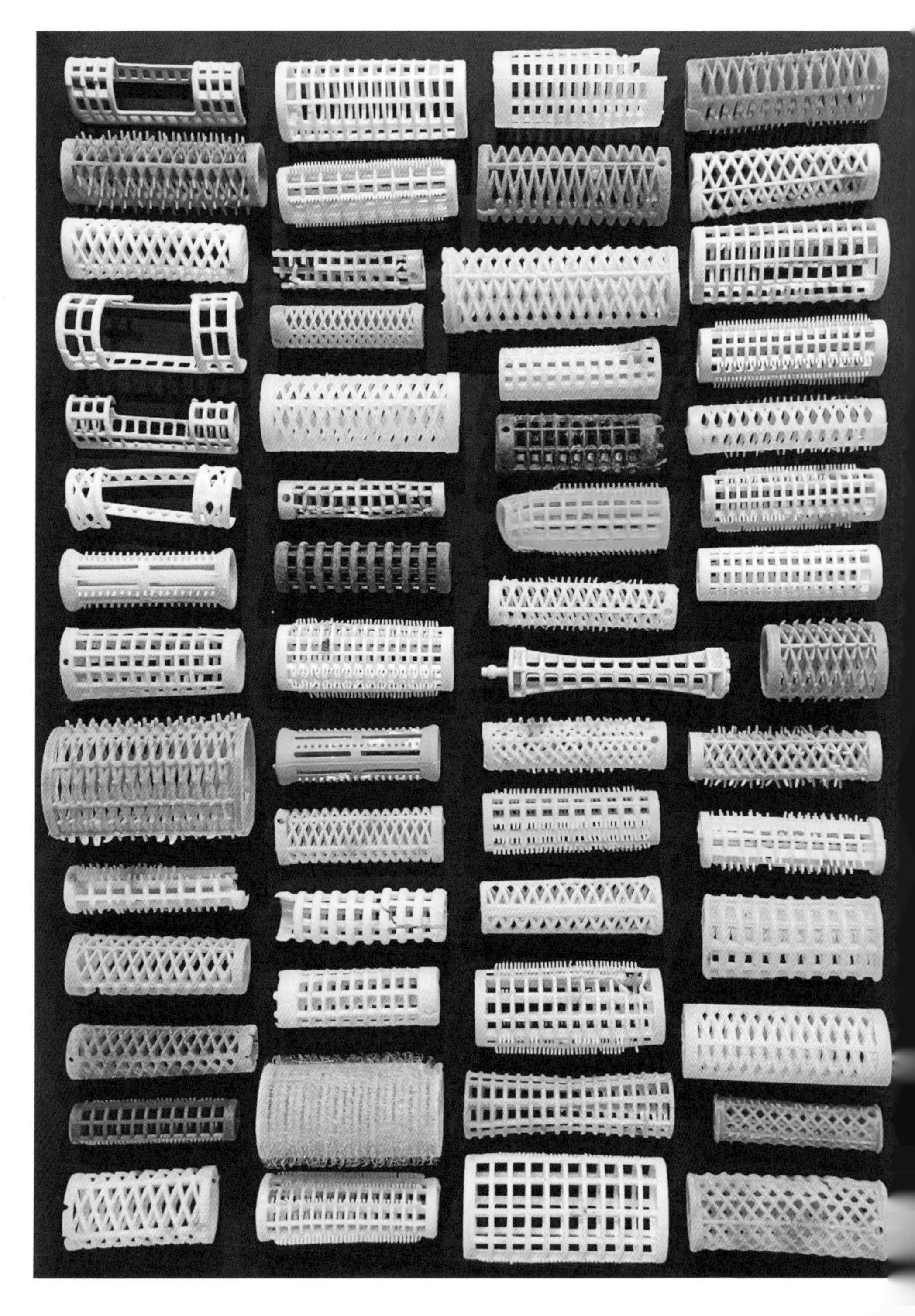

# GLOSSARY

**Adrift:** in or into a drifting condition; so as to drift without being moored, steered, propelled, etc.

**AFOL:** Adult Fan of Lego.

**Bakelite:** the first plastic made from synthetic components, created by the Belgian-American chemist Leo Baekeland in Yonkers, New York, in 1907.

**Beachcomb:** to search for, and collect, objects such as seashells and sea glass along the seashore.

**Beachcombing:** loafing about a port to filch small things (*see* W.H. Smyth and E. Belcher, *The Sailor's Word-Book*, 1867).

**Bio-beads:** tiny plastic pellets sometimes used as part of the filtration process in wastewater or sewage treatment plants. Similar to nurdles but often with a ridged, wrinkled surface.

**Bubble wand:** a device used to form soap bubbles.

**Drift seed:** a seed or fruit adapted for long-distance dispersal by ocean currents. Most come from tropical plants.

**Eyrie:** the nest of a bird of prey, esp. (in later use) that of an eagle on a mountain or cliff.

**Fossick:** a term used in Cornwall, Australia and New Zealand to refer to prospecting for gold, precious stones and fossils. To rummage; to hunt for treasure.

**Gulf Stream:** a great oceanic current of warm water that issues from the Gulf of Mexico and runs parallel to the American coast as far as Newfoundland, and thence in the direction of Europe.

**Ingest:** to take food or liquid (or plastic) into the body.

**Kelp:** a collective term for large brown seaweeds.

**Marine debris:** human-created waste that ends up in oceans, seas and other large bodies of water.

**Mermaid's purse:** the egg-case of a skate, ray or shark.

**Microplastics:** extremely small pieces of plastic, less than 5mm in diameter.

**Minifigure:** also known as a minifig or fig. Small plastic Lego figure measuring 4cm tall.

**Mudlark:** a person who scavenges for interesting or valuable debris in the tidal mud of a river, harbour, etc.

**North Atlantic Drift:** a warm ocean current representing a continuation of the Gulf Stream across the Atlantic, flowing from Newfoundland to the coast of north-west Europe.

**Northeast Passage:** a maritime route through the Arctic along the northern coast of the Eurasian landmass, principally situated off the coast of northern Siberia (Russia).

**Northern fulmar:** a sea bird that frequents the colder waters of the North Atlantic and the North Pacific oceans.

**Nurdles:** small pellets of unprocessed plastic, known too as plastic granulate. Usually smooth and uniform in shape.

**Rogue wave:** a wave of unpredictable size, speed or direction; (now) spec. an exceptionally large wave in the open ocean that far exceeds those encountered in prevailing sea conditions, with heights reaching 25–30m (approx. 82–98ft).

**Sea bean:** beans or seeds of tropical origin, often carried by ocean currents to remote shores.

**Sea glass:** shards of broken glass that have been rolled and tumbled in the ocean until the sharp edges have been smoothed and rounded.

**Spring tide:** an unusually high tide occurring around the time of the new or full moon.

**Strandline:** a mark left by the high tide or a line of seaweed and other debris washed on to the beach by the tide.

**Wreck:** A vessel broken, ruined or totally disabled by being driven on rocks, cast ashore or stranded. Also, that which is cast ashore by the sea in tidal waters; esp. goods or cargo as thrown on land by the sea from a wrecked, stranded or foundered vessel.

# FREQUENTLY ASKED QUESTIONS

***Is there a 'Lego beach'?***
No. Despite what's often written in the press and on social media, there isn't a single 'Lego beach'. Pieces from the spill have been found on many different beaches.

***Where do you find the Lego?***
Beachcombers tend to find the pieces that float on the high tide mark, usually after winter storms and high spring tides. Bricks that sank can occasionally be found among the kelp or brown seaweed that washes ashore. Much of the Lego featured in this book has been discovered by beachcombers and beach cleaners who regularly pick up plastic from beaches, often on a daily basis.

***How do you know if a piece of Lego is from the* Tokio Express?**
Unless it has been hauled up in a fisherman's net off the coast of Cornwall, it can be difficult to tell. Factors to consider are whether the brick was listed on the inventory, where and when it was found, whether similar pieces of Lego or other items from the spill have been discovered nearby and if it's feasible for ocean currents to have carried it to the beach where it drifted ashore. Condition is relevant too. A brick lying in the ocean or buried in sand and sediment for decades will often be more worn than one recently lost.

***Do you know where the shipping containers fell into the water?***
Although we know roughly where the *Tokio Express* lost its containers, not all sank immediately. Some carried on drifting for days.

***Did you identify all one hundred Lego bricks on the inventory?***
All but one.

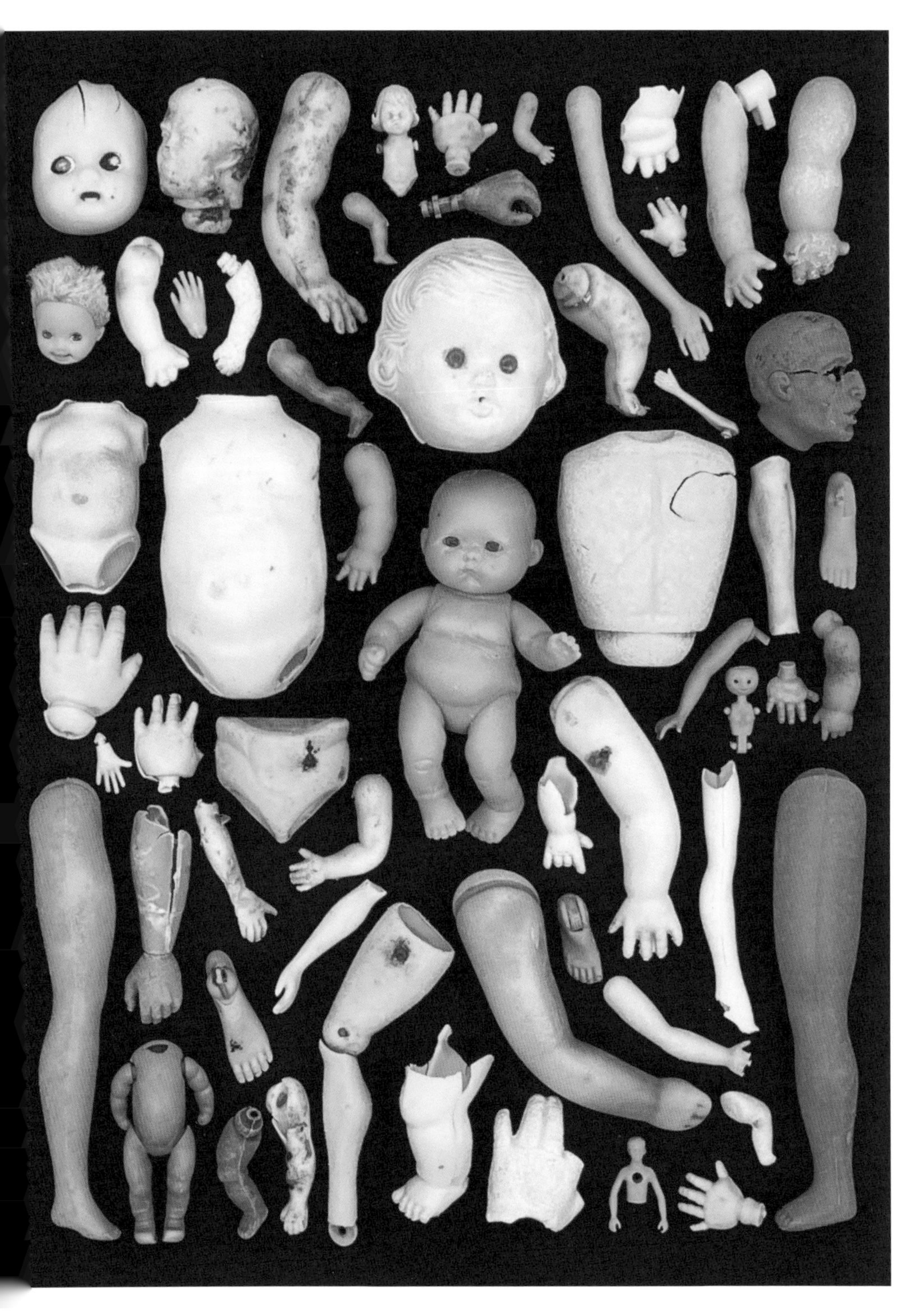

# SELECT BIBLIOGRAPHY

## BOOKS


Cole, P. (2004) *Suspended Animation: an Unauthorised History of Herald & Britain's Plastic Figures*, Plastic Warrior.

Ebbesmeyer, C. and Scigliano, E. (2009) *Flotsametrics and the Floating World*, New York: Collins.

George, R. (2013) *Deep Sea and Foreign Going: Inside Shipping, the Invisible Industry that Brings you 90% of Everything*, London: Portobello Books.

Grigson, G. (1959) *Looking & Finding*, London: Readers Union.

Gunn, C.R. and Dennis, J.V. (1999) *World Guide to Tropical Drift Seeds and Fruits*, Florida: Krieger Publishing Company.

Kemp, C. (2012) *Floating Gold – A Natural (and Unnatural) History of Ambergris*, Chicago and London: University of Chicago Press.

Larn, R. and McBride, D. (1997) *The Cita – Scilly's Own 'Whisky Galore' Wreck*, Cornwall: Shipwreck & Marine.

Nelson, E.C. (2000) *Sea Beans and Nickar Nuts*, London: Botanical Society of the British Isles.

Trewhella, S. and Hatcher, J. (2015) *The Essential Guide to Beachcombing and the Strandline*, Plymouth: Wild Nature Press.

Wallace, J. (1684) *Ane Account off the Ancient and Present State off Orkney*, held by Orkney Library and Archive, reference D101/1.

Woollett, L. (2013) *Sea and Shore Cornwall: common and curious findings*, Zart Books.

Woollett, L. (2016) *Sea Journal*, Zart Books.


## FILMS


*The Wrecking Season* (2004) Directed by Darke, J. (Boatshed Films Ltd)

## SCIENTIFIC PAPERS

Dulvy, N. *et al.* (2021) 'Overfishing drives over one-third of all sharks and rays toward a global extinction crisis', *Current Biology*.

Pétursdóttir, Þ. (2019) 'Anticipated futures? Knowing the heritage of drift matter', *International Journal of Heritage Studies*.

Sheehan, E. *et al.** (2017) 'Strandings of NE Atlantic gorgonians', *Biological Conservation*.

Turner, A. *et al.* (2019) 'Marine pollution from pyroplastics', *Science of the Total Environment*.

Turner, A. *et al.** (2020) 'Weathering and persistence of plastic in the marine environment: Lessons from LEGO, *Environmental Pollution*.

Turner, A. *et al.** (2021) 'Transport, weathering and pollution of plastic from container losses at sea: Observations from a spillage of inkjet cartridges in the North Atlantic Ocean', *Environmental Pollution*.

Turner, A. *et al.** (2021) 'Coastal dunes as a sink and secondary source of marine plastics: A study at Perran Beach, southwest England', *Marine Pollution Bulletin*.

* *Tracey Williams is a co-author.*

## NEWSLETTERS

*Beachcombers' Alert!™*
To subscribe, visit www.flotsametrics.com/beachcombers.php

*The Drifting Seed*
To download back copies, visit www.seabean.com/newsletters

# USEFUL WEBSITES

2 Minute Foundation: www.beachclean.net
Ambergris Connect: www.ambergrisconnect.com
Aphotomarine: www.aphotomarine.com
Blue Planet Society: www.blueplanetsociety.org
British Divers Marine Life Rescue: www.bdmlr.org.uk
Brodie Neill: www.brodieneill.com
Caroline South: www.carolinesouth.co
Cereal Offers: www.cerealoffers.com and
www.youtube.com/user/cerealmad
Cornish Plastic Pollution Coalition: www.cppccornwall.org.uk
Cornwall Wildlife Trust: www.cornwallwildlifetrust.org.uk
Dr Curtis Ebbesmeyer: www.flotsametrics.com
Fishing For Litter: www.fishingforlitter.org.uk
Great Nurdle Hunt: www.nurdlehunt.org.uk
Jo Atherton: www.joatherton.com
Keep Britain Tidy: www.keepbritaintidy.org
The LEGO Group: www.lego.com
Lisa Woollett: www.photographsofthesea.com
Mandy Barker: www.mandy-barker.com
Marine Conservation Society: www.mcsuk.org
Maritime Archaeology Sea Trust: www.thisismast.org
Michelle Costello: www.smartielidsonthebeach.co.uk
Museum of Design in Plastics: www.modip.ac.uk
Nikki Banfield, Barefoot Photographer:
www.facebook.com/barefootphotographer
Ocean Recovery Project: www.keepbritaintidy.org/get-involved/
volunteer/ocean-recovery-project
Plastics Historical Society: www.plastiquarian.com
Rame Peninsula Beach Care: www.ramepbc.org
Receiver of Wreck:
www.gov.uk/government/groups/receiver-of-wreck
Rob Arnold Art: www.instagram.com/rob.arnold.art
Sea Beans and Drift Seeds: www.seabean.com
The Shark Trust: www.sharktrust.org
Surfers Against Sewage: www.sas.org.uk

# ANSWERS

*Page 139 – real pebbles*   *Page 142-43 – plastic shells*

*Page 162-63 Beachcombers' bingo*

1. Jacks from a ball and jacks game
2. Shotgun wads/wadding
3. Zip sliders used in fishing
4. Wheel spacers used in concrete reinforcement
5. Caps from baby food pouches
6. Spokey dokeys (bicycle accessories)
7. Party poppers
8. Discs from toy tracer guns
9. Crab line holder
10. Insulation packs from weather balloons
11. Parts from flying whistle rockets
12. Fish-shaped soy sauce containers
13. Biofilters/biomedia (used in wastewater treatment plants and aquariums)
14. Bubble wands
15. Cap gun ring caps
16. Wearable plastic rings that would once have had a candy jewel attached
17. Twist-off tops from squeezable fruit drink bottles
18. Shotgun cartridges
19. Kite strut holders
20. Firework remnants
21. Balloon accessories
22. Docking rings
23. Widgets from beer cans
24. Poppit beads

*Page 164-65 Beachcombers' bingo 2*

1. Red and yellow bottle caps with ZP logo inside
2. Otrivin nasal decongestant bottles
3. Landmann barbecue wheels
4. Toy roof racks and toy wheels
5. Pink detergent bottles
6. Tommy Hilfiger brand flip-flops
7. Speed brand flip-flops
8. Bunalun Organic rice cake packets
9. Toothbrushes (of type shown)
10. HP printer ink cartridges
11. Hose parts
12. Rolls of tape
13. Toy watches
14. Triangle brand flip-flops
15. Great Wolf Lodge flip-flops
16. Baxter Healthcare intravenous fluid bags
17. Toy truck parts

# PICTURE CREDITS

The people and organisations listed below have kindly given their permission for their images to be reproduced in this book:

Äaron Fabrice de Kisangani: p.49, Lego rigging (France), Lego flipper (Belgium)
André Ellis: p.57, *The Container Ship 'Cita' sinks off St Mary's* illustration
Brodie Neill: p.66, photograph of the 'Gyro' table
Caroline South: p.67, beach finds montage
Chris Easton: p.122, drift card
Dave Ingraham: p.34, Dr Ebbesmeyer with the Friendly Floatees
Delia Webb: p.121, Lego dragons with sea glass
Gilbert Mellaza: p.113, Garfield phones
Giles Richardson: p.76, Lego found on wreck site; p.77, montage of Lego pieces from the spill
Glenys Wynne-Jones: p.19, Margaret Jones
Ian Jepson: p.107, dogfish
Isles of Scilly Museum: p.56, Action Man clothing
Job ten Horn: p.52, Lego dragon
Julie Davies: p.46, Lego cutlass
Liam MacNamara: p.47, Lego dragon (Ireland)
Lisa Woollett: p.45, Lego dragon
Mandy Barker: p.110, printer ink cartridges
Michelle Costello: pp.60-61, Lego found in harbour; pp.104-105 container boxes, cutlass parts, BURPs, radio, log wall
Nikki Banfield: p.46, Lego flipper, Scilly; p.57, trophy part; p.113 Beady Pool beads
Orkney Library and Archive: p.87, page from *Ane Account off the Ancient and Present State off Orkney* (archive reference D101/1)
Rob Arnold: p.64, Lego flippers; p.65, Moai Easter Island head replica, sacks of plastic; p.77, dragon trapped in rocks; p.105, sea-worn dragon
Rosemary Hill: p.47, Lego octopus (Ireland)
Ruth Pickard: p.20, the Polkerris haul
Sam Reoch: p.48, Lego dragon (Guernsey)
Sarah Measham: p.104, remains of Lego octopus
Science Museum Group: p.31, Friendly Floatee duck
Seapixonline.com/Trevor Coppock Photography: p.15, *Tokio Express*
Seth Draper: p.46, Lego dragon
Suki Honey: p.20 Lego dragons; p.24, Lego octopus; p.184 Lego dragon on heart-shaped stone

Terena Hillary: p.104, Lego octopus
Unicorn Publishing Group: cover, pp.19 (dragon), 21 (dragon), 26, 27, 36,37,38, 39 (bat and monster foot), 59, 69, 84, 85 (flowers), 90 (drifter block),96,97,98,99,100 (except dark grey shark),101,102 (light grey shark),125,128 (shoe, bag), 129 (cap, extinguisher, woman),130(plug),141(houses),152,153,154,156,157,164(#1)
Wikimedia Commons: p.86, *Manicaria saccifera*; p.159 Lewis chessman
Wim Kruiswijk: p.50, Lego montage

Endpapers by Jo Atherton
Illustrations and maps on pp.12-13,14,16-17,18,46-49, 51,52,53,56, 79,81,88,94-95,112 by Felicity Price-Smith

All other photographs are the author's own. All natural finds were returned to the shore after photography.

# POETRY CREDITS

'Beachcomber' by George MacKay Brown - permission kindly granted by Hachette UK Ltd.

'maggie and milly and molly and may' by e.e. cummings - copyright 1956, 1984, 1991 by the Trustees for the E.E. Cummings Trust, from *Complete Poems: 1904–1962* by e.e. cummings, edited by George J. Firmage. Used by permission of Liveright Publishing Corporation.

'Sea-Fever' by John Masefield - permission kindly granted by The Society of Authors as the Literary Representative of the Estate of John Masefield.

'Sea Glass' by Bernadette Noll - permission kindly granted by the author.

All other peoms are out of copyright.

# GLOSSARY CREDITS

Adrift, Bubble wand, Eyrie, Gulf Stream, Mermaid's purse, North Atlantic Drift, Rogue wave, Wreck: *Oxford English Dictionary*

Northeast Passage: *Encyclopaedia Britannica*

Beachcomb, Strandline: *Collins English Dictionary*

All other definitions, author's own.

1116211
BONITO DEL NORTE
CFV 100609 FIGGY DUFF
"FOG"

# ACKNOWLEDGEMENTS

First, I would like to thank oceanographer Dr Curtis Ebbesmeyer for his generosity, kindness and for inspiring me to seek out the stories behind the flotsam I find. Without him this book would not have been possible.

My heartfelt thanks also go to former BBC journalist Mario Cacciottolo for shaping the book, editing the script and for buoying me up with good humour when the going got tough.

For his insights, invaluable advice, encouragement and boundless enthusiasm, I am extremely grateful to shipwreck researcher and diver Kevin Heath, who has been immensely supportive throughout.

Huge thanks also go to environmental scientist Dr Andrew Turner at the University of Plymouth, who has shared his knowledge and expertise, and given so much of his time and advice both freely and generously.

I owe a deep debt of gratitude to skipper David Stevens, his brother Alec and father David, who so kindly welcomed me into their world. I can't thank them enough. I am exceedingly grateful, too, to all the other fishermen who shared their finds, stories and pictures, including Bill Brain, Benjamin Haining and Ian Jepson.

I'm particularly indebted to my fellow beachcombers/ beach cleaners and all those who do so much to raise awareness of plastic in the ocean and campaign for a cleaner environment, especially Rob Arnold, Michelle Costello, Delia Webb (for ideas and help with research), Claire Wallerstein, Lisa Woollett (for late night Zoom calls during lockdown to discuss finds), Suki Honey, Terena Hillary, Chris Easton, Lee

James Palmer, Louis-Matisse Nicholls, Sarah Measham, John Collister, Nikki Banfield, Linda Johnston and Mary Woods in England; Seth Draper and Julie Davies in Wales; Rosemary Hill and Liam MacNamara in Ireland; Sam Reoch and Tracy Vibert in the Channel Islands; Aäron Fabrice de Kisangani and Kevin Pierloot in Belgium; Wim Kruiswijk, Jacoline van Duijn and Job ten Horn in the Netherlands; Bo Eide in Norway; Gilbert Mellaza in France; Gui Ribeiro in the Azores; and Ana Pêgo in Portugal. Their observations, insights and suggestions have been invaluable.

Grateful thanks also go to Glenys Wynne-Jones, Ruth Pickard and Gwynneth Bailey for sharing their memories of beachcombing in Cornwall and for permission to reproduce diary entries and the photograph of their dragon-hunting mum Margaret Jones.

Thanks also to Clare Leverton of Fishing for Litter, Neil Hembrow of Keep Britain Tidy, Martin Dorey at #2minutebeachclean and Hugo Tagholm at Surfers Against Sewage.

At the LEGO Group, I'd like to thank Tim Brooks and Therese Noorlander for their time and help, as well as the various designers and employees who have shared their ideas and thoughts over the years.

I am also grateful to fine art photographer Mandy Barker, artists André Ellis, Caroline South and Jo Atherton and furniture designer Brodie Neill for permission to feature their work.

I am deeply indebted, too, to all those who have provided expert information, especially marine biologists David Fenwick and Declan Quigley, marine biologist and beachcomber Dr Paul Gainey, marine ecologist Dr Emma Sheehan, sea bean expert Ed Perry, palaeontologist and 'doctor of decay' Dr Thomas Clements, maritime archaeologist Giles Richardson, archaeologist and researcher Dr Þóra Pétursdóttir at the University of Oslo and Dr Nicholas Melia of the Borthwick Institute for Archives at the University of York.

For a variety of reasons I'd also like to thank John Hourston, Tom Scott, Cornelius Holtorf, professor of archaeology at Linnaeus University, Sweden, Tom Pitchford, Marek Lewcun, Sophie Hawke, Chris and Di Burton, Diana Gorringe, Pam Smethurst, Stuart Baird, Judith Spencer, Andy Dinsdale, Tony Wells, Aaron Bray, Martin Sharp, Jo Carter, Lara Maiklem, Wyl Menmuir, Derek Tait, Maarten Brugge, Martin van der Aa, Paul Morehead of Plastic Warrior magazine, Rob Paton, Dr Max Liboiron, Dr Rebecca Altman, Nick Symes, Neil Redfern, Sarah Maclean and Yvonne Nicoll at Orkney Library and Archive, Philip Morrice, Rose George, Will Stanley, Camilla Moore, Adrian Baldwin, Alison Kentuck, Graham Caldwell, Dr Lucy Hawkes, Dr Huw Lewis-Jones, Andrew Male, Jeremy Batt, Simon Holden, Bart Budding, Kate Hale, curator/manager of the Isles of Scilly Museum and Charlotte Todd at Porthcurno Telegraph Museum.

I'd also like to express my deep gratitude to the immensely talented and patient Felicity Price-Smith at Unicorn Publishing for the beautiful book design and illustrations.

Above all, thank you to Jon, Sophie and Thomas for their love, kindness and for putting up with the ever-growing mounds of flotsam accumulating round the house. I couldn't have done it without you. And a final thank you to Jess, my faithful rescue dog. The best beachcombing companion you could ever wish for.

First published by Unicorn
an imprint of Unicorn Publishing Group LLP, 2022
Charleston Studio
Meadow Business Centre
Lewes BN8 5RW
www.unicornpublishing.org



Unicorn Publishing Group LLP is committed to working with suppliers using FSC sourced paper and other environmentally-friendly sourced products. The logging and manufacturing processes are expected to conform to the environmental regulations of the country of origin.

ISBN 978-1-913491-19-2
10 9 8 7

Design by Felicity Price-Smith
Printed by Fine Tone Ltd

ENDPAPERS
The endpapers have been created by Jo Atherton, an artist who creates striking cyanotype patterns by placing tideline objects from the UK coastline on light sensitive paper. Many of these found objects are plastic, including the highly collectable pieces of Lego from the 1997 container spill. These bold images present an uncanny reflection of ourselves, harnessing the energy upon which we all depend to produce haunting yet familiar remnants of our material culture.